Thermal Analysis and **Design** of **Passive Solar Buildings**

Thermal Analysis and **Design** of **Passive Solar Buildings**

A K Athienitis and M Santamouris

earthscan
from Routledge

First published by James & James (Science Publishers) Ltd. in 2002

This edition published 2013 by Earthscan

For a full list of publications please contact:

Earthscan
2 Park Square, Milton Park, Abingdon, Oxon OX14 4RN
Simultaneously published in the USA and Canada by Earthscan
52 Vanderbilt Avenue, New York, NY 10017, USA

First issued in paperback 2020

Earthscan is an imprint of the Taylor & Francis Group, an informa business

A catalogue record for this book is available from the British Library.

ISBN 13: 978-0-367-57891-6 (pbk)
ISBN 13: 978-1-902916-02-6 (hbk)

Cover photograph: BedZED project by Bill Dunster architects.
Photo by Raf Makda / View Pictures.

Contents

Preface

The term 'passive solar buildings' was originally developed during the early days of solar energy research to distinguish thermal systems in which the primary driving force for solar-induced heat flow is buoyancy (natural convection) as opposed to heat transfer achieved primarily with the aid of fans and pumps.

Generally, if solar energy can be utilized immediately in a building or be directly stored in the thermal mass of the building, without expending a significant amount of energy in transferring heat by mechanical means, then the overall process will be more efficient. In terms of the second law of thermodynamics, it can be shown that the inherent irreversibilities introduced by adding mechanical devices are avoided or minimized.

The term 'passive solar buildings' is employed in this book to emphasize reliance on direct utilization of solar energy in buildings without the use of fans and pumps *as much as possible*. However, in many cases the utilization of fans or pumps is essential in order to transfer stored energy to the controlled space, or to prevent high room temperature swings. Shading devices also need to be effectively controlled to optimize utilization of solar energy, especially its visible component – daylight – while lowering cooling loads.

The 21st century building is rapidly evolving to accommodate three interrelated requirements:

- the achievement of a high quality, easily controllable indoor environment which will enhance comfort and productivity
- minimization of energy costs for processes such as heating, cooling and electric lighting
- sustainable development; one must strive to use environmentally friendly materials with low embodied energy and renewable energy sources such as daylight as much as possible. Furthermore, design solutions must ensure long-term durability of systems, which decrease the impact of waste on the environment.

The area of fenestration in buildings is continuously increasing, driven by two factors – first the higher demand for buildings with much daylight, and second the development of advanced windows with predictable or controlled solar transmittance characteristics (e.g. electrochromic coatings, motorized shading) and high thermal resistance. The increased fenestration areas often result in highly varying heating and cooling loads throughout the year, as well as overheating, especially when inadequate amounts of thermal mass are employed. The fragmented nature of the building process, in which no member of the design team considers the overall optimization of the indoor environment, further compounds the problem. For example, a building designer often implements advanced dynamic elements such as motorized shading devices in the building envelope or airflow windows which are usually not taken into account by the mechanical engineers responsible for heating/cooling system design.

It is the aim of this book to contribute towards the optimization of buildings as systems which act as natural filters between the indoor and outdoor environments, while maximizing the utilization of solar energy.

The book starts with an introduction of basic concepts for passive solar buildings, including a clear day steady-state model and notions of thermal comfort. Chapter 2 overviews essential concepts of heat transfer relevant to passive solar buildings and their dynamic behaviour. Chapter 3 considers analysis of different types of glazing systems, including transparent insulation, shading and daylighting concepts. Chapter 4 introduces several techniques for modelling and analysis of the thermal response of solar buildings, with emphasis on the admittance technique and the explicit finite difference approach; in both cases thermal networks are extensively employed to illustrate the basic modelling concepts. Chapter 5 focuses on the passive response of solar buildings and its analysis through a simple admittance model; this model may be employed to determine room temperature swings and to size thermal storage mass so as to prevent overheating.

Chapter 6 provides an introduction to natural ventilation for good indoor air quality. Chapter 7 presents models for thermal analysis of passive solar buildings intended to help the designer size small auxiliary heating/cooling systems; both convective and radiant systems are considered. Finally, a model of a building with utilization of underground channels for preheating/precooling fresh air for air-conditioning systems is introduced.

Chapter 8 considers in detail thermal comfort and dynamic control of passive solar buildings. Chapter 9 presents several integrative applications: thermal analysis of a building with transparent insulation is introduced in section 9.1, followed by analysis of a solar building with phase-change thermal storage in section 9.2; section 9.3 describes a simple solarium model, and section 9.4 covers photovoltaic systems for buildings, followed by a description of recent advanced fenestration system applications.

CHAPTER 1

Introduction and basic concepts

1.1 Passive solar design principles – an introduction

Solar radiation is the most abundant renewable energy source, without which life on earth would be impossible. It is the driving energy of our ecosystem and of the precipitation cycle. Passive solar design principles have been known to our ancestors since antiquity; for example ancient Romans oriented house openings towards the south so as to be warm in the winter and reduce solar gains in the summer. They also built massive dwellings that reduced room temperature fluctuations. In some countries of the Mediterranean the living room is known as the 'solar room', and in traditional architecture it connects into a south-facing courtyard.

The term 'passive solar building' is a qualitative term describing a building that significantly utilizes solar gains to reduce heating and possibly cooling energy consumption based on natural energy flows – radiation, conduction and natural convection; forced convection based on mechanical means such as pumps and fans is not expected to play a major role in the heat transfer processes. The term 'passive building' is often employed to emphasize utilization of passive energy flows both in heating and cooling.

Passive solar heating systems are generally separated into two broad categories, direct gain and indirect gain (see Section 1.7). When indirect passive systems are insulated from the heated space they are sometimes referred to as isolated.

Passive solar design techniques address the following basic requirements and principles:

- **Transmission and/or absorption of the maximum possible quantity of solar radiation** during winter so as to minimize or reduce to zero the heating energy consumption.
- **Utilization of received solar gains to cover instantaneous heating load and storage of the remainder in embodied thermal mass or specially built thermal storage devices**.
- **Reduction of heat losses** to the environment through use of the appropriate amount of **insulation** and windows with high solar heat gain factor.

- **Shading control devices** or strategically planted deciduous trees to exclude unwanted solar gains, which would create an additional cooling load.
- **Utilization of natural ventilation** to transfer heat from hot zones to cool zones in winter and for natural cooling in the summer; **ground cooling/heating** to transfer heat to/from the deep underground which is at a more or less constant temperature; **evaporative cooling.**
- **Development of integrated building envelope devices** such as windows which include photovoltaic panels as shading devices, or roofs with photovoltaic shingles; the dual role of these elements for electric power production and for exclusion of thermal gains increases their cost-effectiveness.
- Utilization of solar radiation for **daylighting; this requires measures for effective distribution of daylight onto the work plane.**
- **Integration of passive solar systems with the active heating/cooling air-conditioning systems both in the design and operation stages of the building**.

The last requirement is perhaps the most important for the successful design and operation of a building that utilizes passive solar design principles. However, it is usually overlooked because of the absence of collaboration for integration of building design between architects and mechanical engineers. Thus, the architect may often design the building envelope based on qualitative passive solar design principles and the engineer designs the HVAC (heating–ventilation–air conditioning) system based on extreme design conditions, ignoring the benefits due to solar gains and natural cooling. This results in an oversized system, which fights the building rather than using it. The absence of collaboration between the disciplines involved in building design is decreasing with the adoption of computer tools, but the fundamental institutional barriers remain owing to the basic training of architects and engineers which does not foster an integrated design approach.

The design approach proposed in this book is based on the principle that the building and its HVAC system are one thermal system and they must be designed together based on dynamic operation, taking into account thermal storage and control strategy. For example, a variable thermostat setpoint may result in different heating and cooling equipment sizes. Thus, passive solar gains and dynamic building behaviour must be estimated quantitatively under various control strategies to design both the building envelope and the HVAC system properly for harmonious operation.

Depending on climatic conditions and building function, certain heating/cooling systems are more appropriate than others and more compatible with passive systems. For example, the thermal mass in a floor may be used both to store passive solar gains and also for a floor heating system; however, this poses a control challenge which must be carefully considered to achieve acceptable thermal comfort.

A key aspect of passive solar design is choice of the following design parameters:

- **fenestration** area, orientation and type
- amount of **insulation**
- **shading** devices – type, locations and areas
- effective **thermal storage** (insulated from the exterior environment) amount and type (sensible – such as concrete in the building envelope with exterior insulation, or latent – such as phase-change materials).

The above basic design parameters are interlinked and dependent on each other.

The ultimate design objective is minimization of energy costs (heating, cooling, electricity) while maintaining good interior thermal comfort. The thermal mass of the building causes delays in its response to heat sources such as solar gains – the well-known thermal lag effect. This effect, if taken into account in the selection of thermal mass and appropriate control strategies, does not cause thermal comfort problems. It also needs to be taken into account in heating/cooling system sizing. Night ventilation may significantly reduce the need for mechanical cooling.

This book focuses on passive solar systems integrated in the building envelope.

1.1.1 Building enclosure design principles

In implementing passive solar design techniques one must consider all other aspects of building design. A building enclosure and its components should generally be designed to provide a protected and comfortable indoor environment. Building envelope components are built to protect from the weather elements – rain, sun, winds and variations of the environmental temperature. The external envelope acts like a 'filter' between the external environment and the indoor space. The filtering action is best illustrated with the effect of thermal mass on the fluctuations of outside temperature, which are modulated into a small room-temperature swing. Another filtering action is reduction of noise transmission.

In addition to protection from the weather elements there is a need for a structure to support the weight of the building envelope, building contents, snow and water on the roof, and to resist dynamic loads due to wind and earthquakes. In the great majority of buildings the structure is distinct from the enclosure. However, in many cases the structure and envelope functions are combined, such as in shear walls in traditional architecture and in flat roofs reinforced with steel.

This book focuses on thermal and solar performance of the building envelope. It is thus concerned with heat transfer through the building enclosure, including structural components; the thermal interaction between structural elements and building envelope components known as thermal bridge effects often leads to increased heat transfer as well as to differential expansion, which sometimes leads to formation of cracks.

Although the main function of the building external envelope is environmental it has only recently begun to attract the attention that it deserves. It is still relatively common for building designers to place primary emphasis on the visual function, that is, the appearance of a building. Architects, while being well equipped with building designs of traditional types that represent a number of basic models, find it difficult to deal with new concepts in building design such as passive solar techniques and components.

Even though a building envelope can be made of technologically simple components, because of its numerous functions it is still a very complex system. If its operation is not carefully considered throughout the year, a multitude of problems may occur; for example, a passive solar building with large south-facing windows may have a very low heating load, but unless proper shading measures are taken, it may be a 'sauna' in the summer. Furthermore, noise transmission may be another problem, and it should be considered simultaneously with natural ventilation.

Thus, the building envelope is both a barrier and a filter. Sometimes the barrier action is more important, as in shading, while at other times the filtering and modifying action is required. Apart from heat and noise, water must also be excluded from the interior. There are three well-known forms of keeping the interior dry:

- **Permeable barriers** thick enough to permit water that has penetrated the exterior surface to evaporate in dry periods before it reaches the interior surface; this has been the traditional solution for walls.
- **Air gaps**: continuous cavities in the building envelope which break the capillary paths along which water may travel from outside to the interior.
- **Impervious barriers** that prevent transmission of moisture; when they are used special attention to avoiding cracks must be given, otherwise if water penetrates through cracks and is not allowed to evaporate, interstitial condensation may occur.

Heat transmission, heat storage, solar heat gain and infiltration are the principal characteristics of exterior envelopes that influence room temperature and should be carefully considered when determining the effect of building enclosure on indoor climate.

The building shape and the ratio of surface area to volume are major determinants of thermal performance. Early decisions on shape limit future design options on passive solar features such as south-facing window area and natural ventilation strategies. A compact building with small surface area to volume ratio reduces both heat gains and heat losses. However, there are other constraints and factors which often dominate early decisions on form and shape, including:

- land plot shape and size
- building regulations on maximum area to be covered in the plot, maximum height and other urban planning requirements
- shading by adjacent buildings

- daylight needs
- visual and aesthetic factors, needs and preferences of owner
- function of the building and times of operation/occupancy.

The above factors largely constrain our choices on building form and shape. Nevertheless, the designer has several options available to optimize thermal performance, including advanced windows with effective shading mechanisms, natural ventilation strategies and sufficient amounts of thermal mass well distributed to reduce room temperature swings.

A discussion of fundamental considerations on the various building envelope components follows.

EXTERNAL WALLS

Basic decisions about external walls taken at a relatively early stage include the following:

- size, shapes, and position of openings including doors and windows
- treatment of openings, opening arrangement, and protection from heat and water penetration
- construction of solid portions of walls
- combination of passive solar components such as transparent insulation.

Once tentative choices have been made on sizes, positions and arrangements of windows and doors, the solid parts may be designed. A major decision is whether to use heavy or lightweight construction. Heavy construction is preferable for passive solar design and natural cooling in order to reduce room temperature swings. The choice is a function of:

- mass required for heat storage and sound reduction
- degree of fire-resistance required
- desired appearance
- maintenance
- structural, erection and assembly requirements
- availability of materials, cost and constraints.

When a window area is chosen the thermal conductance of the wall as a whole may then be selected. In choosing the different layers of the wall we need to consider the following:

- Water and moisture control. Rain exclusion may be achieved with a cavity in the wall or an impervious barrier. A simple ventilated cavity performs well and is easy to construct. An impervious barrier may be subject to interstitial condensation, and usually requires a vapour barrier and very good joints. Interstitial condensation risk is reduced with a ventilated and drained cavity.

- Thermal control. Methods for achieving the desired U-value include use of rigid thermal insulating sheet or slab materials, low-density insulating block wall materials, cavity wall construction, cavity with filling of insulating materials (plus impervious barrier) and low-emissivity surfaces in cavities.
- Fire control: fire resistance and limitations on combustibility of solid wall materials.
- Incorporation of services (electric, plumbing, heating/cooling, drainage, etc.).

Architects now have a variety of choices available in selecting materials for external walls including the following:

- Masonry: these walls rely on a process of absorption and subsequent evaporation of rainwater close to the surface. Exterior masonry is not useful in storing passive solar gains transmitted through windows.
- Cavity walls: these are designed so that rainwater may penetrate to the inner surface of the outer leaf; a ventilated cavity then allows its evaporation.
- Composite walls: masonry may be used as external skin or backing of external skin. In faced walls facing and backing are bonded so as to result in common action under load. In veneered walls there is no common action.
- Clad masonry walls: single-leaf and solid masonry walls may be used as backing to a cladding system.
- Concrete walls: different types of concrete are available with varying thermophysical properties; *in situ* or precast panels.
- Panels: external cladding panels are manufactured components that speed up construction time – they may be hung on the structure or form part of the structure. Infill panels are supported by the surrounding frames. Note that well-insulated panels exposed to strong sunlight may reach high temperatures (over 100°C), expanding and causing cracks.
- Curtain walls; these are non-load-bearing external walls. They usually consist of a rectangular grid filled with inserts of glass and panels of other materials such as insulation. They are usually suspended in front of a structural frame and their weight plus wind loads are transferred to the frame through point anchors. Curtain walls are increasingly used now in office buildings; special reflective absorptive glasses with shading should be used to reduce cooling loads.

ROOFS

Roofs are a very critical component of the building envelope. They must keep out water and sun, and provide thermal insulation. They must resist the destructive effects of heating and cooling cycles. They are typically of two types: flat and pitched. Flat roofs are protected with an impervious barrier, usually some type of mastic/tar, impregnated felt or metal sheet. Pitched roofs, however, do not have to be totally impervious; overlapping tiles should be backed by lining to prevent entry of wind, wind-driven rain or snow.

Flat roofs in hot climates should be painted with a reflective (e.g. aluminium) paint above waterproofing to reduce radiative gains in the summer, thereby reducing cooling loads. The surface temperature variations are also minimized, reducing the risk of crack formation due to expansion–contraction cycles and subsequent seepage of water.

A pitched roof with a ventilated attic avoids many of the problems of flat roofs. Water drains under gravitational force and heat absorbed by roof tiles (which are free to expand) is lost either to the environment or to the air in the attic space, which is ventilated.

In implementing passive solar systems such as photovoltaic shingles one should always design for proper water drainage and exclude moisture or water.

FLOORS

The lowest floor of a building may be at ground level, below ground level, or elevated on columns. A floor at ground level is the cheapest to build since it requires minimum excavation or fill, and for this reason is the most common in southern Europe. Use of a basement utilizes the stable temperature of the soil to provide a cool environment in the summer and to reduce heat losses in the winter, and for this reason it is popular in extreme climates. A raised floor is often used to provide covered space for parking or other functions, and in some cases for aesthetic reasons.

A floor in contact with the soil must be protected from rising moisture with an impervious barrier or ventilated cavity; the latter, most evident in suspended timber floors, is not as popular today.

Sound transmission through floors is usually not critical, especially for floor slabs, whether suspended or on the ground; the noise will usually be transmitted through windows and doors. A floor is often the best location for thermal mass for storing solar radiation transmitted through a window.

1.1.2 Fenestration and openings in the building envelope

The proper position and size of openings in the building envelope is typically based on experience, judgement, satisfaction of user needs, and building regulations. When passive solar design principles are followed south is the preferred orientation.

Doors are usually chosen based on access needs, security requirements and legislation (e.g. fire protection). One additional consideration in bioclimatic design should be natural ventilation. For example, a door located on the northern side of a house, when opened simultaneously with a south-facing window will induce high rates of airflow by natural ventilation.

Window shapes and sizes are often selected based on preliminary assessment of daylight and view needs; for example, a tall window may provide a view different to that of a wide window with the same area. Sound transmission into a building is usually through windows, especially when they are open. Excessive noise may be reduced with the following measures:

- re-orientation of window wall away from noise source
- use of fixed windows and mechanical ventilation
- use of openable double-glazed windows with 15–20 mm wide cavity
- openable windows with vinyl frame plus rubber lining at the edges, which is airtight when closed.

With the development of scientific techniques for estimating natural ventilation, solar gains and natural daylighting it is now possible to design fenestration systems for specific solar gains in winter, exclusion in summer, natural ventilation and daylighting.

Many modern buildings may be overglazed, sometimes for the sake of appearance, sometimes to achieve high levels of daylighting.

1.2 Solar energy availability

The earth rotates around the sun once a year and at the same time rotates about its polar axis once every 24 hours. During its rotation about the sun it follows an elliptical path with its axis tilted at 23.45° with respect to the plane of the earth's orbit about the sun. On 21 March (Spring Equinox) and 21 September (Autumn Equinox) the sun is directly over the Equator and the length of day and night are equal everywhere on the earth, except the poles. At other times the sun's rays form an angle with the equatorial plane known as the solar declination angle. This angle varies from 23.45° on 21 June to –23.45° on 21 December. All calculations involving solar radiation are based on solar time (ASHRAE 1989).

Solar time is based on the apparent angular motion of the sun across the sky, with solar noon the time that the sun crosses the meridian of the observer. Local standard time (LST) is converted to solar time as follows. First, there is a constant correction for the difference in longitude between the location and the meridian on which the local time is based. Note that 1° in longitude is equivalent to 4 minutes (since 360° is one day). Another correction is the equation of time, ET, which takes into account changes in the earth's rotation.

The apparent solar time (AST) is given by (ASHRAE 1989):

$$AST = LST + ET + 4(LSM - LON) \quad (1.1)$$

where
ET = Equation of Time, minutes
LST = Local Standard Time (positive east of Greenwich)
LSM = Local Standard Time Meridian, degrees
LON = Local Longitude, degrees
4 = minutes of time required for 1° rotation of the earth

A convenient correlation for the equation of time is the following:

$$\begin{aligned} ET = {} & 9.87\sin[4\pi(n - 81)/364] - 7.53\cos[2\pi(n - 81)/364] \\ & - 1.5\sin[2\pi(n - 81)/364] \end{aligned} \tag{1.2}$$

where n = day of year (1–365)

The total solar radiation incident on an inclined surface (Figure 1.1) consists of a direct (beam) part, the sky diffuse solar radiation (e.g. reflected from clouds, and scattered radiation) and the ground-reflected radiation, also assumed to be diffuse.

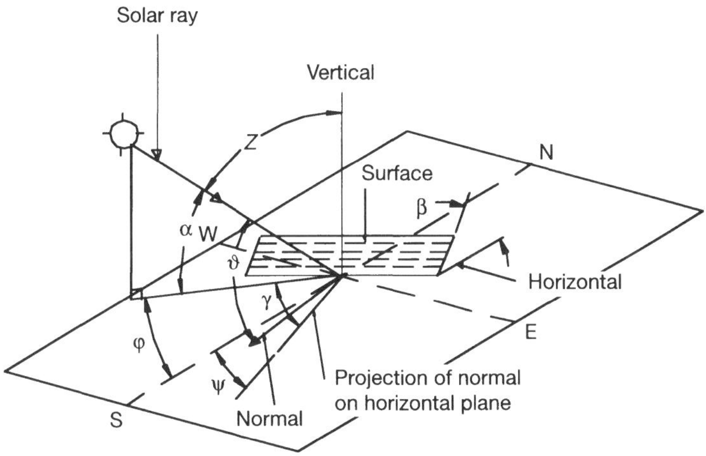

Figure 1.1 Solar radiation geometry for an inclined surface

The position of the sun and the geometric relationships between a plane and the beam solar radiation incident on it may be described in terms of the following angles:

- L latitude, equal to the angle of the location relative to the equator; north is positive
- δ declination, equal to the angular position of the sun at solar noon with respect to the equatorial plane (varies from –23.45 to 23.45°)
- α solar altitude, equal to the angle between the sun's rays and the horizontal (between 0 and 90°)
- z zenith angle, equal to the angle between the sun's rays and the vertical
- φ solar azimuth, equal to the angle between the horizontal projection of the sun's rays from due south (positive in the afternoon)

- γ surface solar azimuth, equal to the angle between the projections of the sun's rays and of the normal to the surface on the horizontal plane
- ψ surface azimuth, equal to the angle between the projection of the normal to the surface on a horizontal plane and due south (east is negative)
- β tilt (slope) angle between the surface and the horizontal (0–180°)
- θ the **angle of incidence** is the angle between the solar rays and a line normal to the surface.

The position of the sun may be expressed as a function of solar altitude and the solar azimuth as shown in Figure 1.1. These angles are a function of the local latitude L and the solar declination δ, which is a function of the date and the apparent solar time (AST) expressed as the hour angle h:

$$h = 0.25 \times (\text{number of minutes from local solar noon})^\circ$$

(h is positive in the afternoon). The declination angle is given by:

$$\delta = 23.45\sin[360(284 + n)/365]^\circ \tag{1.3}$$

The solar altitude, solar azimuth and angle of incidence may be determined by:

$$\sin\alpha = \cos L \times \cos\delta\cos h + \sin L\sin\delta \tag{1.4}$$

and

$$z = 90 - \alpha^\circ$$

$$\cos\varphi = (\sin\alpha\sin L - \sin\delta)/(\cos\alpha\cos L) \tag{1.5}$$

$$\cos\theta = \cos\alpha\cos|\gamma|\sin\beta + \sin\alpha\cos\beta \tag{1.6}$$

where $\gamma = \varphi - \psi$ (note that ψ is negative and φ positive).

Solar radiation is incident on a surface during daylight times (when the hour angle h is less than $|\tan L\tan\delta|$) if the incidence angle is between 0 and 90°.

The total solar radiation I_t incident on a surface is given by the sum of the direct (beam) component I_b, the diffuse sky component I_{ds} and the diffuse solar radiation reflected from the ground I_{dg}.

$$I_t = I_b + I_{ds} + I_{dg} \tag{1.7}$$

where

$$I_b = I_{on}\tau_b\cos\theta$$

with the extraterrestrial solar radiation given by

$$I_{on} = 1353[1 + 0.033\cos(360n/365)] \quad (1.7a)$$

$$I_{ds} = I_{on}\sin\alpha(\tau_d F_{ws}) \quad (1.7b)$$

where the view factor (see Chapter 2) from window to sky is

$$F_{ws} = (1 + \cos\beta)/2$$

$$I_{dg} = I_{on}\sin\alpha(\tau_b + \tau_d)\rho F_{wg} \quad (1.7.c)$$

where the view factor from window to ground is

$$F_{wg} = (1 - \cos\beta)/2$$

where τ_b and τ_d are the beam and diffuse atmospheric transmittance respectively and ρ is ground reflectivity. Typical values of ground reflectivity are given in Table 1.1. There are more accurate anisotropic models for diffuse solar radiation (Klucher 1979) that may be employed.

Table 1.1 Solar reflectivity of various surfaces (Threlkeld 1970)

Surface	*Incidence angle (°)*					
	20	*30*	*40*	*50*	*60*	*70*
New concrete	0.31	0.31	0.32	0.32	0.33	0.34
Bright green grass	0.21	0.22	0.23	0.25	0.28	0.31
Bitumen and gravel roof	0.14	0.14	0.14	0.14	0.14	0.14
Bituminous parking lot	0.09	0.09	0.10	0.10	0.11	0.12
Crushed rock	0.20	0.20	0.20	0.20	0.20	0.20

For any transparent layer, part of the incident radiation (I) is transmitted (τI), another portion is reflected (ρI) and the remainder is absorbed (αI), that is

$$\alpha + \rho + \tau = 1 \quad (1.8)$$

where
α = absorptance
ρ = reflectance
τ = transmittance

For grey bodies we assume that the radiation properties do not depend on wavelength. Usually the grey body model is applied over a specific wavelength range, and we determine averaged properties over this range. Details on computation of radiation properties are given in Chapter 2.

Example

Determine the local solar azimuth and altitude at 9.30 Central Time on 23 October at 35°N latitude and 95°W longitude. Also, determine the incidence angle for a vertical surface facing southeast.

First we determine ET, then AST and hour angle h:

$$n = 273 + 23 = 296$$

$$\mathrm{ET}(296) = 16.115 \text{ min}$$

$$\mathrm{LON} = 95°$$

$$L = 35°$$

$$\mathrm{AST} = 9.5 + 16.115 + (90 - 95)/15 = 9.435 \text{ h}$$

$$h = (\mathrm{AST} - 12) \times 15 = -38.47°$$

Knowing these basic quantities we can then determine the position of the sun and the angle of incidence.

$$\delta = 23.45\sin[360(284 + n)/365] = -12.446°$$

$$\alpha = \arcsin[\cos(35)\cos(-12.446)\cos(-38.47) + \sin(35)\sin(-12.446)] = 30.175°$$

$$\varphi = \arccos\{[\sin(30.175)\sin(35) - \sin(-12.446)]/[\cos(30.175)\cos(35)]\} = -44.646$$

Note that the solar azimuth is negative because it is east of south. We now determine the incidence angle:

$$\beta = 90°$$

$$\psi = -45° \text{ (negative – east of south)}$$

$$\gamma = \varphi - \psi = 0.354°$$

$$\theta = \arccos[\cos(30.175)\cos(0.354)\sin(90) + \sin(30.175)\cos(90)] = 30.117°$$

The **sunrise and sunset times** may be determined by evaluating the times at which the incidence angle (or zenith angle) on a horizontal surface is equal to zero. The sunset hour angle h_s is given by:

$$h_s = \arccos[-\tan(L)\tan(\delta)] \tag{1.9}$$

Similarly, $-h_s$ is the sunrise angle. The corresponding times are determined by dividing by 15°/hour.

1.2.1 Hottel's clear sky model

A convenient method developed by Hottel (1976) is employed here to determine the beam radiation transmitted through a clear atmosphere. The atmospheric transmittance for beam radiation is given by:

$$\tau_b = a_0 + a_1 \exp[-k/\cos(z)] \quad (1.10)$$

where the constants a, k depend on climate and altitude A (km).

$$a_0 = r_0[0.4327 - 0.008\ 21(6 - A)^2]$$

$$a_1 = r_1(0.5055 - 0.005\ 95(6.5 - A)^2)$$

$$k = r_k(0.2711 + 0.018\ 58(2.5 - A)^2)$$

The correction constants for climate types are given in Table 1.2.

Table 1.2 Correction factors for Hottel's clear sky model (Hottel 1976)

Climate type	r_0	r_1	r_k
Tropical	0.95	0.98	1.02
Midlatitude summer	0.97	0.99	1.02
Subarctic summer	0.99	0.99	1.01
Midlatitude winter	1.03	1.01	1.00

After determining the beam transmittance (τ_b) of the atmosphere, we may determine also the diffuse transmittance (Liu and Jordan 1960):

$$\tau_d = 0.271 - 0.294\tau_b \quad (1.11)$$

The total daily extraterrestrial solar radiation H_{oh} on a horizontal surface may be determined by integration of the instantaneous radiation between sunset and sunrise times. For an inclined surface we also need to determine the sunset and sunrise hour angles on the surface, which may be different from the daily sunset and sunrise hour angles:

$$h_s' = \min\{\arccos[-\tan(L)\tan(\delta)], \arccos[-\tan(L - \beta)\tan(\delta)]\} \quad (1.12)$$

where min indicates the smaller of the two terms in the brackets.

Note that the solar radiation incident on a south-facing surface is symmetric about solar noon. Therefore, the total daily solar radiation incident on a south-facing surface is determined by integrating for a half day and multiplying by 2.

Example: Evaluation of solar radiation incident on an inclined surface

Determine the collector tilt angle for which the total daily solar radiation incident on a south-facing collector located at a latitude of 45°N is maximum on 21 January and for 21 March.

We need to numerically integrate the instantaneous solar radiation evaluated at several time intervals:

$\beta = 0, 10, 20...90°$ $\psi = 0$ for south-facing surface

Assume $\rho = 0.2$ for ground reflectivity

$\delta = -20.14°$ for 21 January and 0 for 21 March

The sunrise time was found and divided into eight equal time intervals.

$$t_s' = h_s'/15$$
$$= \min\{\arccos[-\tan(L)\tan(\delta)], \arccos[-\tan(L - \beta)\tan(\delta)]/15\} = 4.56 \text{ h}$$

Time interval $\Delta t = t_s'/8$

For each value of tilt angle β and time t, the incidence angle was determined and then the beam and diffuse incident radiation were calculated. The atmospheric beam transmittance and beam radiation incident on a vertical surface (21 January) are shown in Figure 1.2. The total solar radiation as a function of tilt angle is given in Figure 1.3 for 21 January and 21 March.

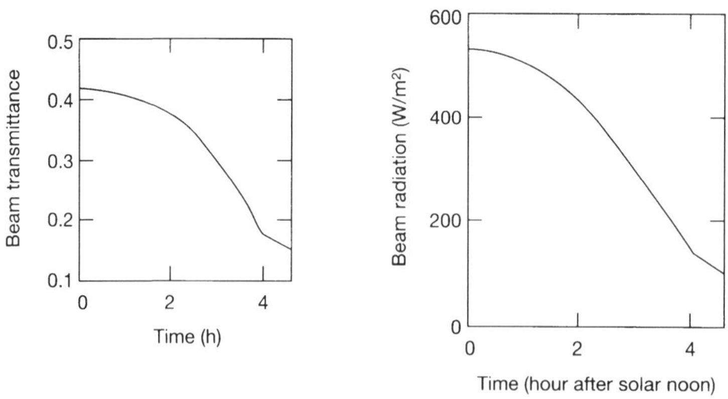

Figure 1.2 Beam transmittance of atmosphere and incident solar radiation on vertical surface for 21 January

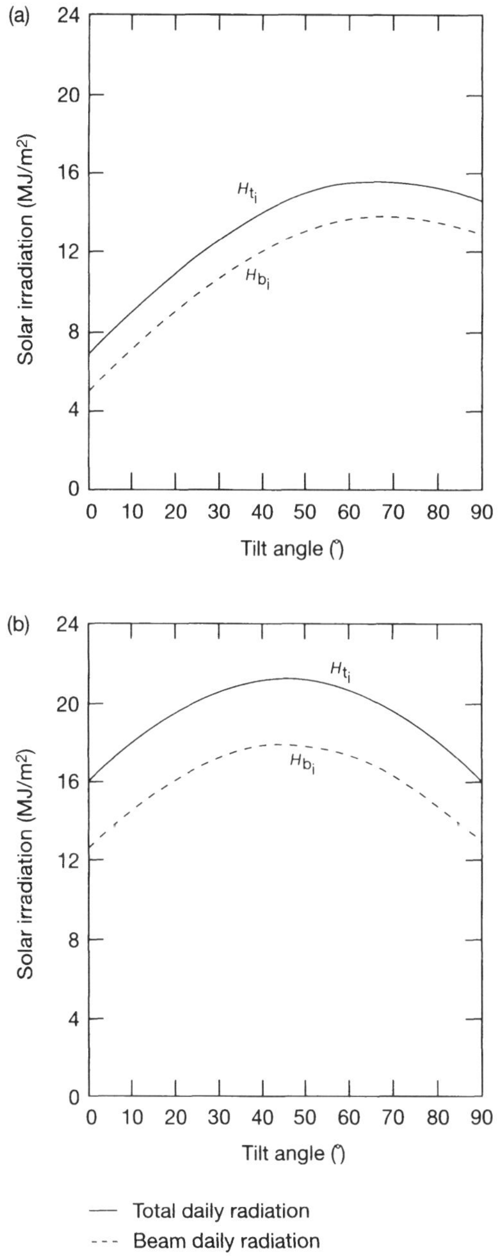

Figure 1.3 Total solar radiation and beam radiation incident on south-facing surface as a function of tilt angle for (a) January and (b) March

As can be seen from Figure 1.3 the optimum tilt angle for this latitude for January is approximately 65° with a peak quantity of about 16 MJ, as opposed to 45° and 22 MJ for March. The difference between each pair of curves is the amount of diffuse solar radiation incident on the surface (from the sky plus ground-reflected). Note that the vertical surface is not significantly inferior to the optimally tilted surface for January (receives about 1.5 MJ less).

Figures 1.4 and 1.5 show the solar radiation incident on vertical surfaces of various orientations for latitudes 30°N, 40°N, 50°N and 60°N on typical clear summer days.

1.3 Microclimate and buildings

Climate is the average of the atmospheric conditions over an extended period of time over a large region. Small-scale patterns of climate, resulting from the influence of topography, soil structure, ground and urban forms, are known as **microclimates**.

The local environment around buildings may significantly alter the effective environment temperature 'felt' by the building (Honjo and Takakura 1990). For example, trees and lawns may lower the temperature in the summer by a few degrees compared with asphalt or concrete surfaces. A densely populated city centre such as Los Angeles (Akbari *et al.* 1992) may experience local outdoor temperatures as much as 8°C higher than green suburban areas.

The principal parameters characterizing climate are: air temperature, solar radiation, humidity, and precipitation and wind. The climate of cities differs from the climate of surrounding rural areas mainly due to the structure of cities and heat released by vehicles. In general, the climate in cities is characterized by high ambient temperatures, reduced relative humidity, reduced wind speed and reduced received direct solar radiation.

As a consequence of heat balance, air temperatures in densely built urban areas are higher than the temperatures of the surrounding rural country. The phenomenon, known as 'heat island', is due to many factors, the more important of which are summarized by Oke *et al.* (1991) and deal with: (a) the canyon radiative geometry that contributes to decrease the long-wave radiation loss from within street canyons due to the complex exchange between buildings and the screening of the skyline, (b) the thermal properties of materials that increase storage of sensible heat in the fabric of the city, (c) the anthropogenic heat released from combustion of fuels and animal metabolism, (d) the urban greenhouse, which contributes to increase the incoming long-wave radiation from the polluted and warmer urban atmosphere, (e) the canyon radiative geometry decreasing the effective albedo of the system because of the multiple reflection of short-wave radiation between the canyon surfaces, (f) the reduction of evaporating surfaces in the city putting more energy into sensible and less into latent heat, and (g) the reduced turbulent transfer of heat from within streets.

Urban heat island studies refer usually to the 'urban heat island intensity', which

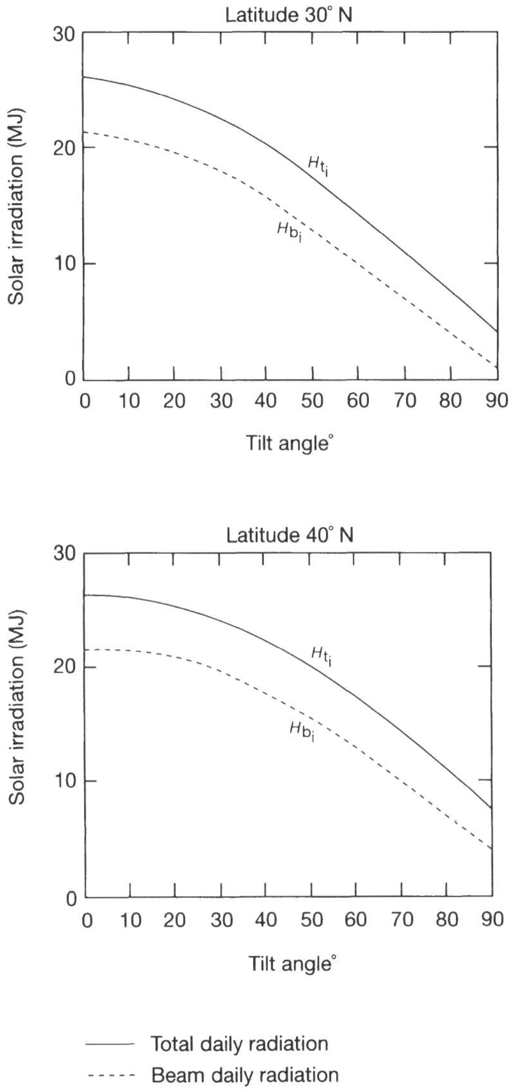

Figure 1.4 Total solar radiation incident on south-facing inclined surface (L = 30°N, 40°N) for various tilt angles (30 June)

is the maximum temperature difference between the city and the surrounding area. Data compiled by various sources (Santamouris 2001a) show that heat island intensity can be as high as 15°C. Extensive studies on the heat island intensity in Athens involving more than 30 urban stations show that urban stations present temperatures higher than reference suburban stations by between 5 and 15°C.

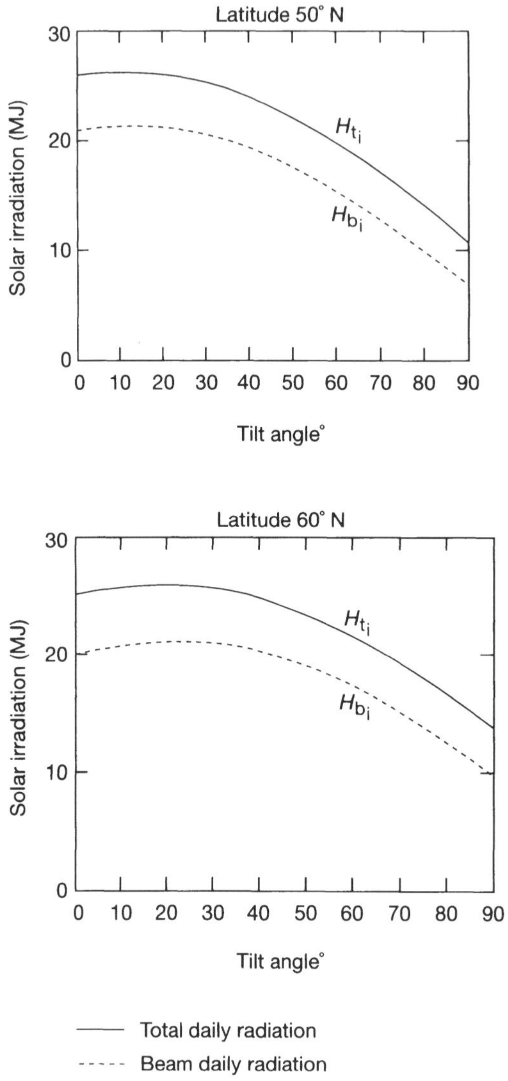

Figure 1.5 Total solar radiation incident on south-facing inclined surface (L = 50°N, 60°N) for various tilt angles (30 June)

In addition to the temperature increase, the urban environment affects many other climatological parameters. Global solar radiation is seriously reduced because of increased scattering and absorption. As mentioned by Landsberg (1991), the sunshine duration in industrial cities is reduced by 10–20% in comparison with the surrounding countryside, and similar losses are observed in received energy.

In a general way, wind speed and direction in the canopy layer are seriously decreased compared with the undisturbed wind speed. This is mainly due to the specific roughness of a city, channelling effects through canyons, and also because of the heat island effect.

In parallel, the urban environment affects precipitation and cloud cover. The exact effect of urbanization depends on the relative place of a specific city regarding the general atmospheric circulation. In Budapest, mainly because of industrialization, cloud cover has been increased by 3% during the winter period. As mentioned by Escourrou (1991), urbanization causes a proportional increase in precipitation in cities such as London, which because of its geographical location is more often in the zone of perturbations than in cities such as Paris. Studies in Bombay, India, have shown that the development of an industrial zone close to the city has increased precipitation by 15% (Rao and Murty 1973).

Addressing successful solutions to counterbalance the effects of temperature increase on the energy performance of urban buildings is an important priority. Successful solutions to counterbalance the effects of the urban microclimate involve the use of more appropriate materials, increased use of green areas, use of cool sinks for heat dissipation, appropriate layout of urban canopies, etc.

The optical characteristics of materials used in the outdoor facade of buildings or in streets and pavements and especially the albedo for solar radiation and long-wave emissivity have a very important impact on the urban energy balance and the performance of buildings. Increase of the surface albedo has a direct impact on the energy balance of a building. Large-scale changes on urban albedo may have important indirect effects on the city scale. Numerous studies have been performed to evaluate direct effects from albedo change and it has been concluded that important energy gains must be expected (Santamouris 2001a).

Vegetation modifies the microclimate and the energy use of buildings by providing solar protection, lowering air and surface temperatures and increasing the relative humidity of the air because of evapotranspiration. Trees also help mitigate the greenhouse effect, filter pollutants, mask noise, prevent erosion and calm their human observers. Water surfaces modify the microclimate of the surrounding area, reducing the ambient air temperature, either by evaporation or convection. Fountains, ponds, streams, waterfalls or mist sprays may be used as cooling sources, for lowering the temperature of the outdoor air and of the air entering the building.

Akbari *et al.* (1992) present the results of computer simulations aimed at studying the combined effect of shading and evapotranspiration of vegetation on the energy use of typical one-storey buildings in various US cities. They found that by adding one tree per house, the cooling energy savings range from 12% to 24%, while adding three trees per house can reduce the cooling load between 17% and 57%. According to this study, the direct effects of shading account for only 10–35% of the total cooling energy savings. The remaining savings result from temperatures lowered by evapotranspiration.

The presence of trees or buildings around a building may also significantly alter the wind environment around it. This should be taken into account in designing for enhancement of natural ventilation in the summer. It is important to orient certain openings (windows/doors) towards the prevailing wind direction and to design for some windows in the leeward direction so that the wind may easily flow through the building.

Shading of a building by surrounding buildings or trees should be taken into account when estimating solar gains for summer cooling calculations and in locating and designing passive solar components such as a solarium or an attached sunspace. A deciduous tree may be ideal in providing summer shading in front of a sunspace while letting the winter sun inside. In the case of office buildings daylighting considerations may be even more important.

Natural evaporative cooling may significantly reduce the temperature of outdoor air entering a building for natural ventilation in the summer. This may be achieved by air flowing over an outdoor pond and fountain. The saturation efficiency is an important factor in identifying the cooling potential for all evaporative cooling systems:

$$\text{Saturation (humidifying) efficiency} = (T_{db,in} - T_{db,out})/(T_{db,in} - T_{wb,in})$$

1.4 Steady-state thermal analysis of buildings

The basic steady-state heat flow (q) equation is based on Fourier's conduction law:

$$q = -Ak(\mathrm{d}T/\mathrm{d}x) \tag{1.13}$$

where

k = thermal conductivity

A = surface area through which heat flows

T = temperature

x = distance

L = thickness

Written in discrete (finite difference) form the above equation becomes:

$$q = -Ak(T_{x=L} - T_{x=0})/L \tag{1.14}$$

$$q = (T_{x=0} - T_{x=L})/R = \Delta/R$$

where thermal resistance of wall:

$$R = L/(kA)$$

The other basic steady-state equation describes heat flow based on a heat transfer coefficient h, which may represent convection (h_c), radiation (h_r) or an approximate combined coefficient, often called a film coefficient h (equal to $h_c + h_r$).

$$q = Ah(T_s - T_a) \tag{1.15}$$

Steady-state thermal analysis techniques are often employed to estimate peak heating loads for preliminary sizing of heating systems. For this purpose the following equation is used to estimate the maximum heating load q_{max}:

$$q_{max} = (U_{tot} + U_{inf})(T_{room} - T_{o\_min}) \tag{1.16}$$

where

T_{room} = room thermostat setpoint

T_{o_min} = minimum exterior design temperature

U_{tot} = sum of all building envelope element conductances (windows, walls, doors, etc.)

U_{inf} = conductance due to infiltration based on the following constitutive equation:

$$\text{Infiltration heat loss} = q_{inf} = \dot{m}C_{p,air}(T_R - T_o) = U_{inf}(T_{room} - T_{o\_min})$$

$$U_{inf} = \dot{m}C_p = \text{number of air changes per second} \times \text{Vol}_{room} \times (\rho C_{p,air})$$

where $\dot{m}$ = mass flow rate

The thermal resistance of a building envelope component is determined as the sum of the thermal resistance of each layer plus the resistance of the interior and exterior film coefficients (h_i and h_o respectively):

$$R = 1/h_i + 1/h_o + \Sigma_i(L_i/k_i) \tag{1.17}$$

where L_i and k_i are thickness and thermal conductivity of layer i. For an interior cavity, its thermal resistance is also added to the above equation. Typical heat transfer coefficients for various surfaces and cavities are computed in Chapter 2.

For a pitched roof with pitch angle b, a void and a horizontal ceiling, the total thermal resistance per unit area of ceiling is given by

$$R = R_r\cos(b) + R_v + R_c \tag{1.18}$$

where R_r, R_v and R_c are the thermal resistances of the roof, void and ceiling per unit area respectively.

In most building envelope components such as walls we have heat flow paths that cross structural support members such as wood studs in wood frame construction, concrete beams and columns in concrete structures, frames in windows and so on.

Computation of the effective thermal resistance of wall assemblies may be based on either of two simplifying assumptions:

- Parallel heat flow paths, one being through the wall area containing structural members and the other one through the remaining area without structural members. In this case we have two thermal resistances in parallel.
- Isothermal planes at different wall layer surfaces.

The actual value of the wall thermal resistance is expected to lie between the values given by the two models. The first model is usually closer to reality and is typically used. The actual wall thermal resistance may be determined using controlled tests, particularly the guarded hot box test (ASTM standard 1989), or *in situ* measurements (Norlen and Bloem 1994). It may also be determined with the two- or three-dimensional finite difference thermal bridge model (see Chapter 4).

In determining the thermal resistance of a wall with fraction of area containing structural members equal to A_s, the effective thermal resistance per unit area R_{eff} is determined by assuming parallel heat flow paths as follows:

$$R_{eff} = 1/[(A_s/R_s) + (1 - A_s)/R_i)] \qquad (1.19)$$

where:
A_s = wall surface area where heat flow crosses structural members
R_s = thermal resistance (per unit area) through path A_s
R_i = thermal resistance (per unit area) through path without structural members

Heat flow through solid ground floors is generally of a two- or three-dimensional nature and it may be accurately analysed with numerical models or approximate analytical solutions.

1.5 Principles of thermal comfort

The ultimate aim of the building thermal design process should be to ensure good thermal comfort under most operating conditions and weather patterns. There exist numeral comfort models for thermal comfort such as Fanger's predicted mean vote (PMV) model (ASHRAE 1989) and other models more appropriate for summer conditions (Athienitis *et al.* 1995). A person's feeling of thermal comfort is a function of air temperature T_{ai}, mean radiant T_{mr} at the location of interest, air movement and humidity, and also personal factors such as clothing and activity.

Various environmental indices are often used as a measure of thermal comfort, including equivalent temperature, effective temperature, humid operative temperature, globe temperature and (in the UK) the resultant temperature. Several environmental parameters that affect thermal comfort can be measured directly:

- air temperature T_{ai}
- wet-bulb temperature T_{wb}
- dew-point temperature T_{dp}
- water vapour pressure p_v
- total atmospheric pressure p
- relative humidity RH
- specific humidity W
- air velocity V
- mean radiant temperature T_{mr}; the temperatures of individual surfaces are usually combined into T_{mr}

The mean radiant temperature is a very important variable in all thermal comfort calculations. It is the uniform temperature of an imaginary black enclosure in which radiant heat transfer from a person equals the radiant heat transfer in the actual enclosure.

T_{mr} may be estimated from measurements of globe temperature, air temperature and air velocity. T_{mr} can also be calculated from measured temperatures of each room interior surface. Since building surfaces have a high long-wave emittance, the following blackbody approximation for T_{mr} is satisfactory:

$$T_{mr}^4 = \Sigma_i(T_i^4 \times F_{p-i}) \tag{1.20}$$

where

i = 1..N indicates room surface

T_i = temperature of surface i, k

F_{p-i} = view factor between a person and surface i

If small temperature differences exist between the room surfaces, then the above equation can be expressed by its linear form:

$$T_{mr} = \Sigma_i T_i \times F_{p-i} \tag{1.21}$$

In a rectangular room one possible approximation in calculating the view factors is to model a person as a sphere, for ease of calculation (this would be a reasonable approximation for a seated person). The mean radiant temperature may also be calculated from the plane radiant temperature in six directions (ASHRAE 1989).

The **operative temperature** is defined as the uniform temperature of an enclosure in which an occupant would exchange the same amount of heat by radiation plus convection as in the actual non-uniform environment (ASHRAE 1989); it is given by

$$T_e = (h_r T_{mr} + h_c T_{ai})/(h_r + h_c) \tag{1.22}$$

where T_{ai} is the air temperature, T_{mr} is the mean radiant temperature, and h_r and h_c are radiative and convective coefficients, respectively, for a person or object (sensor). The operative temperature is a particularly important parameter for passive solar design because the mean radiant temperature is usually higher in well-insulated passive solar buildings. In a poorly insulated building with inefficient windows the mean radiant temperature in the winter will tend to be lower than that in an energy-efficient, well-insulated building for the same air temperature. Thus, to achieve the same thermal comfort in a poorly insulated building we need to achieve the same operative temperature by raising the air temperature to compensate for the lower mean radiant temperature.

One other factor that may be important in thermal comfort calculations for passive solar buildings is the effect of direct and diffuse solar radiation incident on people; it can raise the operative temperature felt by as much as 10°C (Athienitis and Haghighat 1992).

Example

Consider the room shown in Figure 1.6. For a wall *R*-value equal to 4 RSI (thermal resistance, m^2°C/W), a window *R*-value equal to 0.3 RSI and an outside temperature of −10°C, determine the operative temperature for a room air temperature equal to 20°C. Assume room dimensions 3 × 3 × 3 m^3, window 2 × 2 m^2, infiltration one air change per hour, film coefficient on interior surfaces 10 W/m^2°C.

$$R_w = 0.3 \text{ RSI}$$

$$R_{wall} = 4 \text{ RSI}$$

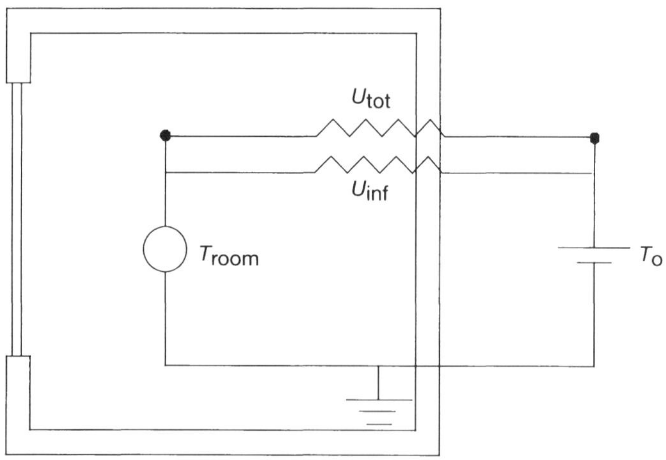

Figure 1.6 Simple steady-state model of heat transfer through building

$T_o = -10°C$

$T_{ai} = 20°C$

$h = 10 \text{ W/m}^2°C$

Infiltration conductance:

$$U_{inf} = \text{Vol} \times (\text{ach}/3600) \times \rho c_{air} = 27(1/3600)1200 = 9 \text{ W/°C}$$

Envelope conductance:

$$U_{tot} = (A_w/R_w) + (A_{wall}/R_{wall}) = 4/0.3 + (3 \times 3 \times 6 - 4)/4 = 25.8 \text{ W/°C}$$

Temperatures:

$$T_w = T_{ai} - (T_{ai} - T_o)(A_w/R_w)/(A_w h) = 20 - (30/hR_w) = 10°C$$

$$T_{wall} = T_{air} - (T_{ai} - T_o)(A_{wall}/R_{wall})/(A_{wall} h) = 20 - (30/hR_{wall}) = 19.25°C$$

$$T_{mr} = (A_w T_w + A_{wall} T_{wall})/A_{total} = 18.56°C$$

$$T_e = T_{ai} + 2T_{mr} = 19.05°C$$

A more detailed discussion of thermal comfort is presented in Section 8.1.

1.6 Approximate estimation of solar savings

An approximate estimation of energy savings due to utilization of solar gains during the heating season can be made by means of a steady-state analysis using the techniques described above, including the mean monthly value of the transmitted solar radiation. Considering the simple zone model in Figure 1.6, and including now the mean transmitted solar radiation for the month G_0, we may perform an energy balance at the room interior node to obtain the resulting mean room temperature without heating as follows:

$$G_0 = (U_{tot} + U_{inf})(T_{room} - T_{o\_mean}) \tag{1.23}$$

Therefore we obtain

$$T_{room} = T_{o\_mean} + G_0/(U_{tot} + U_{inf}) \tag{1.24}$$

The solar savings as a fraction of the total auxiliary heating load (with no solar gains) are then given by the ratio:

$$\text{Solar fraction} = (T_{room} - T_{o\_mean})/(T_{setpoint} - T_{o\_mean}) \tag{1.25}$$

The above technique gives a very preliminary approximate estimation of solar savings, without taking into account overheating and the effect of thermal mass. However, despite its shortcomings, this estimate can provide the designer with a good estimate of the potential building thermal performance as a function of window type, area, orientation and amount of insulation.

More detailed dynamic thermal analysis is necessary to size the heating and cooling systems properly as well as for more accurate analysis of building energetic performance under different control strategies such as a lower thermostat setpoint at night.

Example

Consider the room shown in Figure 1.6. Determine the percentage solar savings for the room, assuming that the outside temperature is 0°C and the mean daily solar gains for December are 12MJ/m² of window.

$$T_{room} = 0°C + (A_w \times 12000000 \text{ J})/[86400 \text{ s} (25.8 + 9)] = 12°C$$

$$\text{Solar fraction} = (12 - 0)/(20 - 0) = 0.6 = 60\%$$

1.7 Conceptual design of passive solar buildings

One of the basic decisions in building design concerns form and shape. Generally basic passive solar principles dictate a rectangular form with the longest side facing south or near-south. Another basic decision is the type of passive solar system to be selected for each zone of the building, type of fenestration, thermal storage and insulation. The main types of passive solar systems are shown in Figure 1.7 and discussed below.

Direct gain systems rely primarily on south-facing windows (north-facing in the southern hemisphere) and thermal mass distributed on room interior surfaces. The room acts essentially as both thermal collector and thermal storage. The key decisions to be made are choice of window area and type and quantity of thermal mass. A dynamic thermal analysis is necessary to ensure that the ratio of peak solar gains to thermal mass does not exceed the maximum room temperature swing set by thermal comfort considerations.

The major advantage of direct gain systems is their simplicity: that solar energy is directly absorbed by the thermal storage or released by convection into the room air. Thus, no heat transfer devices are required, with their inherent inefficiencies, and temperatures remain relatively low, resulting in reduced heat losses. Well-designed shading systems are required to reduce unwanted solar gains in the summer, but to allow maximum gains in winter. Because they are the simplest to design and build, direct gain systems are also the most popular.

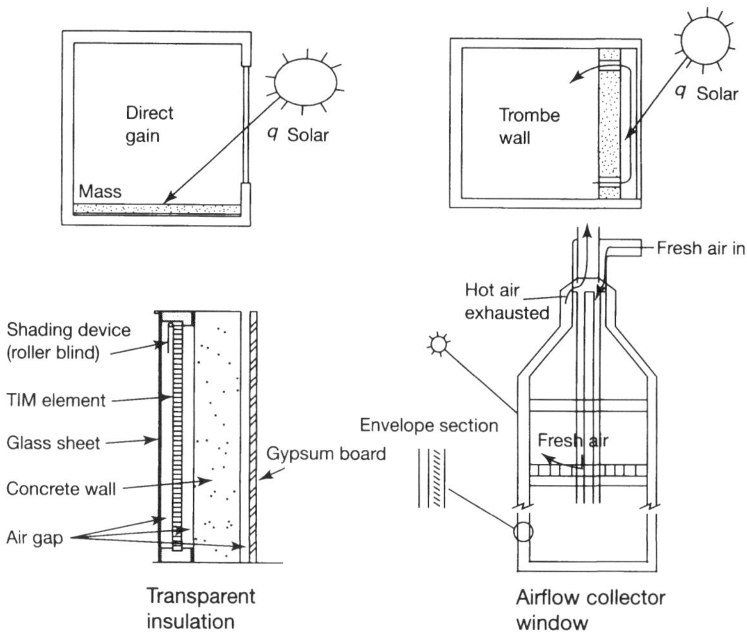

Figure 1.7 Main types of passive solar systems integrated in the building envelope

Systems in which the thermal storage mass is separate from the main building envelope systems are known as **indirect gain systems**. Of these, the most common is the collector-storage wall, also known as the mass or Trombe wall.

- **Collector-storage wall (Trombe wall) systems:** part of the south-facing wall is glazed and inside the glazing there is a thick layer of thermal mass such as concrete, brick or water tanks with a black surface to absorb and store the transmitted solar radiation. It may have vents at the top and bottom to permit transfer of some of the stored heat by convection from the face across the glazing, which may reach a high temperature (e.g. 70°C). Most of the stored heat is transferred by convection and radiation from the interior surface to the room interior, but a portion is lost through the glazing or glazings to the exterior. Night insulation and high thermal resistance fenestration systems reduce this loss.
- **Transparent insulation systems:** the same principle as for the storage walls is employed but in this case the solar energy collection and storage is achieved in an exterior wall. Instead of having glazings in front of the thermal mass, in this case we have transparent or translucent insulation such as a plastic honeycomb with relatively high thermal resistance.
- **Air heating systems such as air-flow windows and air collectors:** in air-flow windows, a dark surface such as dark blinds between two glazings is employed

to absorb the solar heat, which then rises by natural convection, and assisted by the fan to flow through hollow floors/ceiling mass or an isolated thermal storage device such as a rock bed. The principle is very similar to **air flow solar collectors**, installed in the past on the roof of a building. By integrating air flow collectors into the building envelope, the overall building cost is reduced.

- **Solar chimneys** are often employed for natural cooling. In these systems, one wall of the chimney is heated by solar radiation, warming up the air, which rises by buoyancy to the outside, creating a negative pressure and drawing cooler outside air into the building. Use of thermal mass keeps the solar chimney warm at night so that this beneficial natural cooling process continues during night-time when the outside air is even cooler. If the air is passed over water we may also enhance this process through evaporative cooling. The thermosyphon effect may also be employed with water-heating solar collectors for domestic hot water. In this case, the collector is usually a separate mechanical device not integrated in the building envelope.

1.8 Bibliography

Akbari H., Davis S., Dorsano S., Huang J. and Winett S. (1992) *Cooling our Communities – A Guidebook on Tree Planting and Light Colored Surfacing*, US Environmental Protection Agency, Office of Policy Analysis, Climate Change Division.

ASHRAE (1989) *ASHRAE Handbook – Fundamentals*, Atlanta, Georgia, USA.

ASTM (1989) Guarded Hot Box Test (C236), *ASTM Book of Standards*, ASTM, USA.

Athienitis A.K. (1990) 'A method and algorithm for estimation of transmitted solar radiation and its long-term averages using Fourier series', *Solar Energy*, **45**:5, 257–263.

Athienitis A.K. (1994) *Building Thermal Analysis*, Electronic Mathcad Book, MathSoft Inc., Boston, USA.

Athienitis A.K. and Haghighat F. (1992) 'A study of the effects of solar radiation on the indoor environment', *ASHRAE Transactions*, **98**:1, 257–261.

Athienitis A.K., Berger X. and Santamouris M. (1995) *Thermal Analysis of Buildings for Summer Comfort*, European Commission Directorate General for Energy DG XVII Publication.

CIBSE (1986) *CIBSE Guide*, Volume A, CIBSE, London.

Escourrou G. (1991) *Le Climat et La Ville*, Nathan Université, Paris.

Fanger P.O. (1970) *Thermal Comfort*, McGraw Hill, New York.

Gianccone A. and Gianfranco R. (1986) 'The influence of solar radiation distribution in a room on thermal comfort evaluation', *Proc. CIB*, 3185–3192.

Honjo T. and Takakura T. (1990) 'Simulation of thermal effects of urban green areas on their surrounding areas', *Energy and Buildings*, **15**:3–4, 443–446.

Hottel H.C. (1976) 'A simple model of transmittance of direct solar radiation through clear atmospheres', *Solar Energy*, **18**.

Klucher T.M. (1979) 'Evaluation of models to predict insolation on tilted surfaces', *Solar Energy*, **23**, 111–114.

Landsberg H.E. (1981) *The Urban Climate*, Academic Press.

Liu B. and Jordan R. (1960) 'The interrelationship and characteristic distribution of direct, diffuse and total solar radiation', *Solar Energy*, **4**, 1–19.

Norlen U. and Bloem J.J. (1994) 'Comparison of methods for thermal resistance estimation', European Conf. on Energy Performance and Indoor Climate in Buildings, Lyon, France, Nov., pp. 804–809.

Oke T.R, Johnson G.T., Steyn D.G. and Watson I.D. (1991) 'Simulation of surface urban heat islands under "ideal" conditions at night – Part 2: diagnosis and causation', *Boundary Layer Meteorology*, **56**, 339–358.

Rao T.R. and Murty Bh. V.R. (1973) 'Effect of steel mills on rainfall at distantly located stations', *Indian Journal of Meteorology and Geophysics*, **24**:1, 15–26.

Santamouris M. (2001a) *Energy and Climate in the Urban Built Environment*, James and James (Science Publishers) Ltd, London.

Santamouris M. (2001b) 'Solar and natural resources for a better built environment', *Solar Energy, the State of the Art*, James and James (Science Publishers) Ltd, London.

Threlkeld J.L. (1970) *Thermal Environmental Engineering*, 2nd edn, Prentice Hall, Englewood Cliffs, NJ.

CHAPTER 2

Transient heat transfer and thermal storage

2.1 Transient heat conduction

Conduction is the process of heat transfer in which heat is transferred due to the presence of a temperature gradient in a body without macroscopic movement of the molecules. Molecules within a high-temperature region have a higher internal energy, and through microscopic agitation transfer some of their energy to adjacent 'cooler' molecules. At a macroscopic (not molecular) level, the heat flow was first described by Fourier's conduction law.

In the x, y, z directions, we have the following heat flows:

$$q_x = -k_x \partial T/\partial x \tag{2.1}$$
$$q_y = -k_y \partial T/\partial y$$
$$q_z = -k_z \partial T/\partial z$$

where k_x, k_y and k_z are the thermal conductivities (W/mK) in the three directions. For most materials, which are isotropic, the thermal conductivity is equal in all directions.

Now considering a control volume of dimensions dx dy dz, and applying an energy balance over a time interval dt we obtain the three-dimensional heat flow equation:

$$\rho c \partial T/\partial t = k_x \partial^2 T/\partial x^2 + k_y \partial^2 T/\partial y^2 + k_z \partial^2 T/\partial z^2 + S \tag{2.2}$$

where S is an internal heat source (W/m^3), c is specific heat capacity and ρ density.

The thermal capacity of a building, most of it due to thermally massive materials – so-called thermal mass – stores heat during the day and modulates room temperature swings. Solar radiation incident on the exterior building envelope as well as high outside temperature cause heat flow into the building interior. Solar radiation absorbed by a wall and outdoor temperature may be combined into an equivalent temperature known as sol-air temperature T_{eo} (ASHRAE 1989).

Part of the solar radiation incident on the wall outer surface is absorbed while the remainder is reflected. If the wall is covered with transparent insulation, then part of the incident solar radiation is also transmitted. A portion of the absorbed solar radiation plus heat gain from outside air is conducted into the wall; part of this conduction heat is stored in the solid wall layers, raising their temperature while the remainder reaches the room interior surfaces. Hotter surfaces exhibit a net loss of radiant heat (long wave) to colder room interior surfaces, as well as heat loss by convection to the room air, raising its temperature.

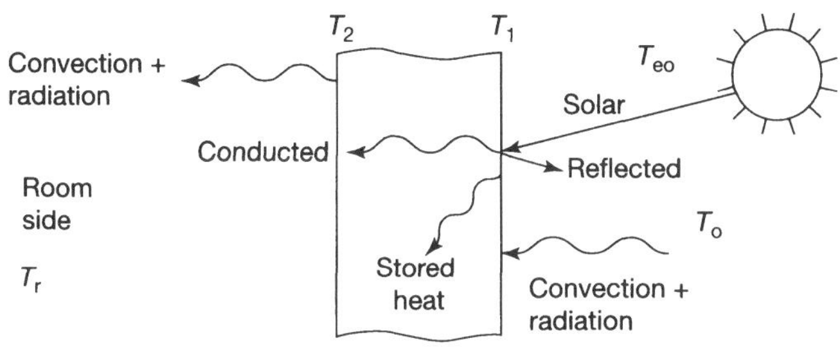

Figure 2.1 Heat exchange mechanisms for a layer of thermal mass

The heat transfer process for the wall in Figure 2.1 is governed by the following parabolic, diffusion-type partial differential equation:

$$\partial T/\partial t = \alpha \partial^2 T/\partial x^2 \tag{2.3}$$

with boundary conditions ($x = 0$ corresponds to exterior surface)
(i) at $x = 0$

$$q = -k\partial T/\partial x = G + h_o(T_o - T_1) = h_o(T_{eo} - T_1) \tag{2.3a}$$

where the sol-air temperature, $T_{eo} = T_o + \alpha_s G/h_o$
(ii) at $x = L$

$$q = h_i(T_2 - T_r) \tag{2.3b}$$

where α is thermal diffusivity, k thermal conductivity, α_s solar absorptance, q heat flux and h heat transfer coefficient.

One-dimensional conduction through the walls is generally assumed, as well as uniform irradiation of their surfaces by solar radiation. However, when thermal bridges are present, such as at corners in rooms and for heat loss through the floor, two-dimensional or three-dimensional analysis is required.

2.1.1 Special cases of one-dimensional heat transfer with analytical solution

HEAT CONDUCTION IN SEMI-INFINITE SLAB

Heat conduction in a semi-infinite slab is the closest one-dimensional model for heat transfer to the ground. Although not an exact representation of real situations, it can provide preliminary estimates which may be used to validate more detailed numerical models.

A semi-infinite slab is a model for a body with a single plane surface ($x = 0$) and its other surfaces distant enough to ignore for time periods of interest in transient analysis. If a uniform boundary condition is applied at $x = 0$, it is reasonable to assume that this case can be analysed as transient one-dimensional conduction. One case that closely fits this model is the ground, with a uniform surface or air temperature; if we measure the soil temperature deep into the ground away from the surface, one would expect that the temperature is not significantly affected by what is happening at the surface. We have three main boundary conditions:

Case 1: Specified temperature T_s at $x = 0$ imposed at time $t = 0$ with initial uniform body temperature T_i. The solution is given by:

$$T(x,t) = T_s + (T_i - T_s) \times \mathrm{erf}[x/2(\alpha t)^{1/2}] \tag{2.4}$$

Case 2: Specified heat flux q_s at $x = 0$ with initial temperature T_i:

$$T(x) = \frac{2q_s}{k}\sqrt{\frac{\alpha t}{\pi}}\exp\left(\frac{-x^2}{4\alpha t}\right) - \frac{q_s x}{k}\left(1 - \mathrm{erf}\left(\frac{x}{2\sqrt{\alpha t}}\right)\right) + T_i \tag{2.5}$$

The relation between surface temperature T_s, initial temperature and heat flux is given by

$$q_s = (T_s - T_i)/2(\alpha t/\pi)^{1/2}$$

Case 3: Convective boundary condition h at $x = 0$, environment temperature T_e and initial temperature T_i:

$$T(x,t) = (T_e - T_i)\left[1 - \mathrm{erf}\left(\frac{x}{2\sqrt{\alpha t}}\right) - \exp\left(\frac{hx}{k} + \frac{h^2\alpha t}{k^2}\right)\left(1 - \mathrm{erf}\left(\frac{x}{2\sqrt{\alpha t}} + \frac{h\sqrt{\alpha t}}{k}\right)\right)\right] + T_i \tag{2.6}$$

The heat flow at the surface at time t is given by

$$q_s = h(T_e - T_i) \times \exp(h^2\alpha t/k^2) \times \{1 - \mathrm{erf}[h(\alpha t)^{1/2}/k]\} \tag{2.7}$$

Example

The ground is initially at uniform temperature 20°C. If it is then subjected to a temperature of −10°C, determine the depth at which a pipe should be laid to prevent freezing.

This is case 1 of the semi-infinite model. Of course, the ground is never at a uniform initial temperature; however, for design purposes, this is an adequate model to locate the safe depth for placing the pipe to avoid freezing.

Assume soil properties:

ρ = 2000 kg/m^3
c = 840 J/kg K
k = 0.9 W/m K
$\alpha = k/\rho c = 5.37 \times 10^{-7}$ m^2/s
t = 60 days × 86 400 s/day

$$T(x,t) = -10 + (20 - -10) \times \text{erf}[x/2(\alpha t)^{1/2}] \quad (2.4)$$

Figure 2.2 shows the temperature variation with depth. As can be seen, the depth at which the temperature is 0°C after 60 days is about 1 m.

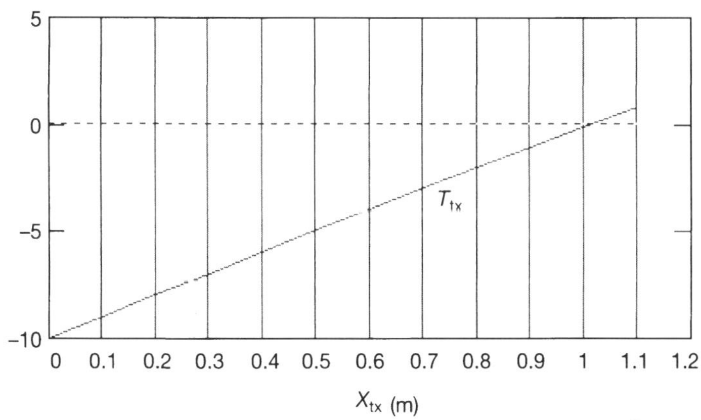

Figure 2.2 Temperature variation with depth for example (case 1)

Example

A high-intensity infrared ceiling heating system is employed to keep a factory floor warm. If the heating system is switched on with the floor initially at temperature 10°C, determine the radiant heat flux intensity on the floor required to warm the floor surface to a temperature of 20°C after time t. What is the floor temperature at the bottom surface of the concrete slab?

Assume that the concrete floor slab is 20cm thick and that the semi-infinite model case 2 applies.

Concrete properties:

$\rho = 2200\ \text{kg/m}^3$

$c = 840\ \text{J/kg K}$

$k = 1.7\ \text{W/m K}$

$\alpha = k/\rho c = 9.66 \times 10^{-7}\ \text{m}^2/\text{s}$

$$q_s = (20 - 10)k/2(\alpha t/\pi)^{1/2} = 147.5\text{W/m}^2$$

Now, we will determine the temperature as a function of depth using equation 2.5 with $t = 3$ hours.

Examination of the temperature variation (Figure 2.3) with depth at $t = 3$ hours shows that the heat flux on the surface has no significant effect below a depth of 0.2m. Therefore, the present model is satisfactory. Note that if there was a significant heating of the lower boundary of the slab then heat loss through this boundary would have to be included in the analysis.

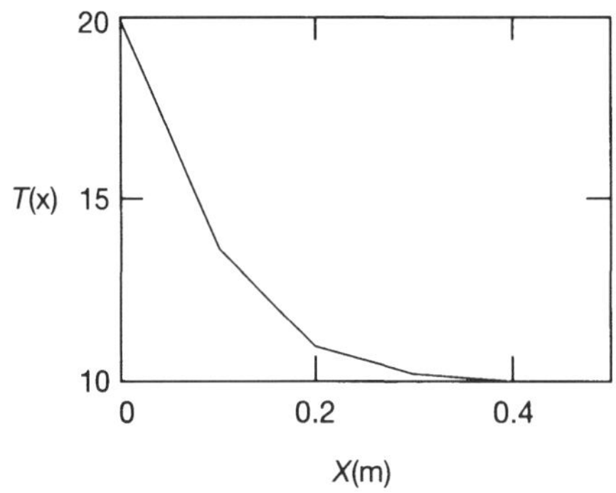

Figure 2.3 Temperature variation with depth for example (case 2)

Example

A floor slab made of concrete 30cm thick is initially at a uniform temperature 20°C. Suddenly, one of its surfaces is subjected to convective cooling with a heat transfer coefficient of $14\text{W/m}^2\text{K}$ into an environment at 30°C. Calculate the temperature at various depths from the surface at time t after the change.

Concrete properties:

$\rho = 2200\ \text{kg/m}^3$

$c = 840\ \text{J/kg K}$

$k = 1.7\ \text{W/m K}$

$\alpha = k/\rho c = 9.66 \times 10^{-7}\ m^2/s$
$h = 14\ W/m^2\ K$
$T_e = 30°C$
$T_i = 20°C$

Substituting in equation 2.6 and calculating the temperature at two depths (surface and bottom) we obtain the results in Figure 2.4.

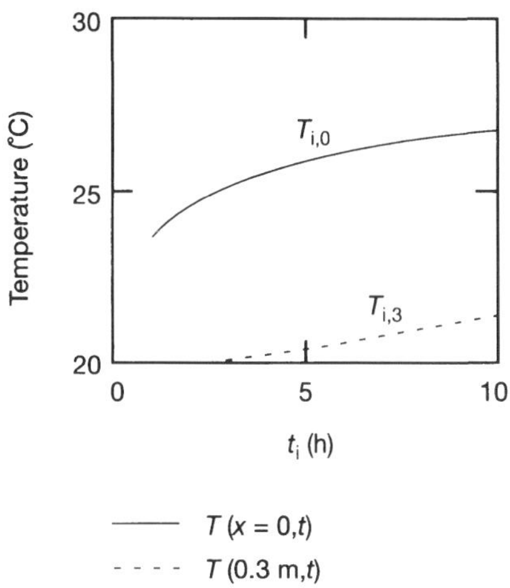

Figure 2.4 Temperature variation with time for two depths ($x = 0$ m, $x = 0.3$ m) for example (case 3 of semi-infinite solid–convective boundary condition)

WALL WITH INTERNAL HEAT GENERATION

Heat may be generated within a wall. For example, electric radiant panels, which often form the interior layer of a ceiling or a wall to provide radiant heat, contain electric resistance elements that generate heat. This heat can be approximated as internal heat generation.

We often may assume one-dimensional heat conduction. If a steady-state analysis is performed then the relevant energy balance is given by the following ordinary differential equation:

$$kd^2T/dx^2 + Q = 0 \tag{2.8}$$

where k is the thermal conductivity of the wall layer and Q is the rate of internal heat generation. Consider, for example, a radiant panel of area $1m^2$ and thickness 13mm made of gypsum board with electric resistance elements built into it with a total power output of 100W assumed to be uniformly generated within the panel (Figure 2.5).

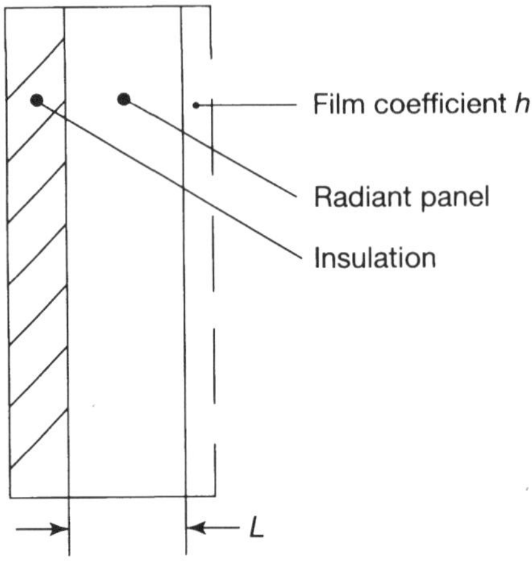

Figure 2.5 Radiant panel built into wall

Boundary conditions:
1. At $x = 0$ – adiabatic: $dT/dx = 0$
2. At $x = L$ – convective: $-k dT/dx = h(T - T_{room})$

Solution:
$T(x) = T_{room} + Q(L/h - (L^2 - x^2)/2k)$
For the particular case of gypsum board (Figure 2.6) assume
$L = 13$ mm, $k = 0.16$ W/m K, $T_{room} = 20°C$, $h = 12$ W/m^2°C

CONDUCTION SHAPE FACTORS

Conduction shape factors are convenient parameters for expressing the effect of geometry in two-dimensional heat transfer problems, usually involving a source and a sink. The conduction shape factor S is defined based on the relationship:

$$q = kSD_T \tag{2.9}$$

where q is heat flow, k the thermal conductivity and D_T the temperature difference between source and sink. The factors S have been determined for various situations with analytical techniques. One useful case is that of a pipe buried in soil.

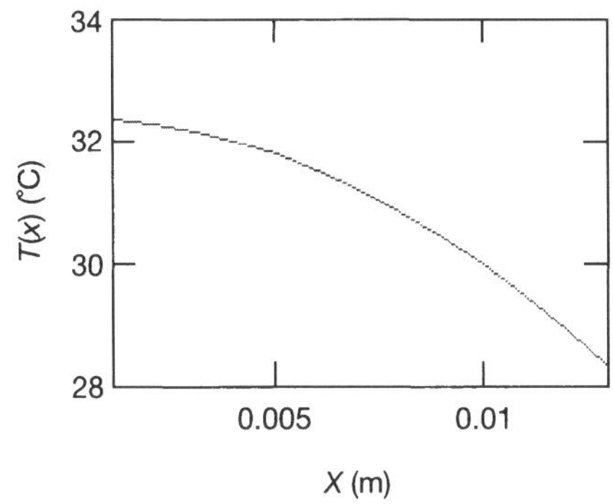

Figure 2.6 Temperature variation with distance from room surface for gypsum panel

Example

Consider a long pipe of diameter D, length L and temperature T_p buried horizontally at a depth z in soil with constant surface temperature T_s.

L = 20 m, D = 0.2 m, T_s = 0°C, T_p = 70°C, k = 0.7 W/m K

The variation of heat loss from the pipe with depth at which it is buried is given in Figure 2.7.

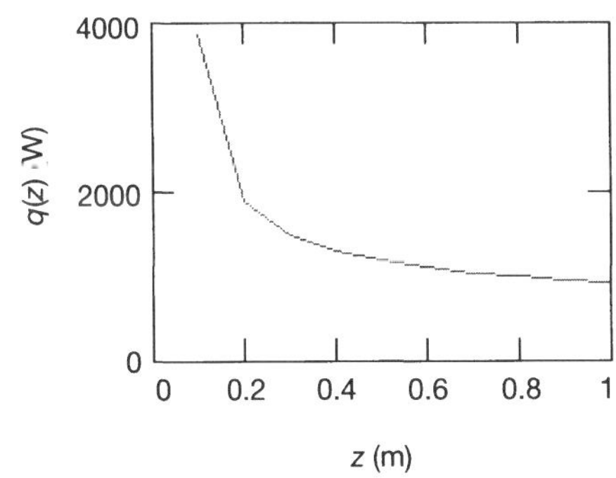

Figure 2.7 Variation of heat loss from isothermal pipe buried at depth z horizontally

2.2 Natural and forced convection

When a fluid is locally heated or cooled its density changes relative to that of the surrounding fluid and a buoyancy force is created which causes the warmer and lighter fluid to rise; cooler parts of the surrounding fluid descend and a natural circulation pattern is created. This process is observed in a direct gain passive solar zone as shown in Figure 2.8.

The floor, which is heated by incoming solar radiation, warms the air, which rises along the back wall and the ceiling and then descends along the cooler window surface – getting cooled until it reaches the floor. This process is known as **natural convection**. It is the basic process for transfer of heat by fluids in passive solar buildings.

Usually heat transfer by convection is described by a linearized heat transfer coefficient h_c:

$$q = Ah_c(T_w - T_{ai}) \tag{2.10}$$

where T_w and T_{ai} are wall and air temperatures respectively. Before considering simplified relationships for h_c for air, which can be linearized to obtain constant h_c values, we will consider a simple form of the general (Navier–Stokes) fluid flow equations for two-dimensional natural and forced convection.

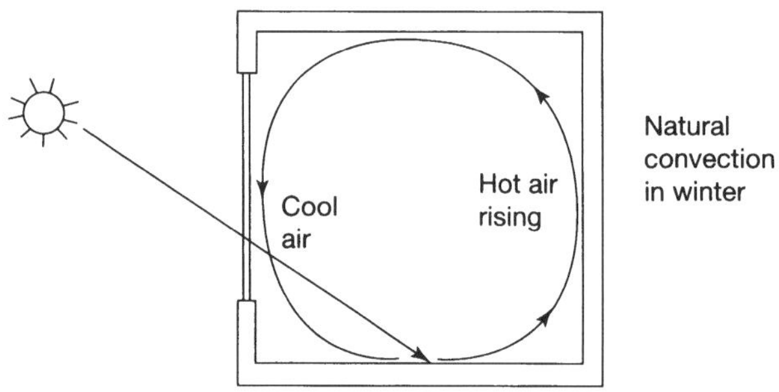

Figure 2.8 Natural convection in a direct gain room in winter

Forced convection is the fluid flow caused by an external force or device, which imposes a free stream velocity U_0. The basic differential equations that describe fluid flow are the continuity (conservation of mass), momentum and energy equations.

These are given below in dimensionless form as follows for two dimensions for a boundary layer over a plate of length L (in the x direction and free-stream flow U_0 in the same direction).

Continuity, which expresses the conservation of mass:

$$\frac{\partial U}{\partial X} + \frac{\partial V}{\partial Y} = 0 \tag{2.11}$$

where $U = u/U_0$, $X = x/L$, $Y = y/L$, $V = v/U_0$ with u and v being the velocity components in x and y directions, U_0 the free stream velocity.

Momentum:

$$U\frac{\partial U}{\partial X}+V\frac{\partial V}{\partial Y}=-\frac{dP}{dX}+\frac{1}{Re}\frac{\partial^2 U}{\partial Y^2} \tag{2.12a}$$

where $P = p/(\rho/U_0^2)$.

Energy conservation:

$$U\frac{\partial T}{\partial X}+V\frac{\partial T}{\partial Y}=\frac{1}{Re\cdot Pr}\frac{\partial^2 T}{\partial Y^2} \tag{2.12b}$$

where T is dimensionless temperature $T = (T' - T_0)/(T_s - T_0)$; T' = fluid temperature at (x,y), T_s = plate temperature and T_0 is free stream temperature.

In the case of combined natural and forced convection over a plate the momentum equation becomes (with the addition of a buoyancy term):

$$U\frac{\partial U}{\partial X}+V\frac{\partial V}{\partial Y}=-\frac{dP}{dX}+\frac{1}{Re}\frac{\partial^2 U}{\partial Y^2}+\frac{Gr}{Re^2}T \tag{2.12c}$$

The dimensionless numbers employed in the above equations are:

Re = Reynolds number = U_0L/ν (ratio of inertia and viscous forces)
Pr = Prandtl number = ν/α = kinematic viscosity/thermal diffusivity
Gr = Grashof number = $g\beta\Delta TL^3/(\nu^2)$ = ratio of buoyancy to viscous forces where $\beta = 1/T$, expansion coefficient for air (approximated as ideal gas) at temperature T, K
Ra = Rayleigh number = $g\beta\Delta TL^3/(\nu\alpha)$ = $Gr\,Pr$
Nu = Nusselt number = hL/k = dimensionless heat transfer coefficient (for $Nu = 1$ we have pure conduction across an air layer with $h = L/k$).

Heat flow relationships that apply to different fluids are often obtained experimentally and then expressed as correlations of the dimensionless numbers discussed above. Based on these general correlations we develop specific equations that apply to airflow in buildings, air heating devices, etc. These equations are often linearized and approximate representative values of heat transfer coefficients are obtained for use in linear thermal system analysis.

When a detailed analysis of heat and fluid flow in 3D is desired the usual approach is discretization in space and time using the control volume method. Numerical techniques are extensively applied to passive solar analysis in Chapter 3; numerical and spatial discretization is usually applied to conduction heat transfer because of its dominant role in energy storage, thermal lag effect and room temperature fluctuation. Detailed numerical analysis of convection heat transfer is not usually necessary for building design purposes.

2.2.1 Natural convection in wall cavities and windows

Cavities (air spaces) exist in walls, windows, etc. The unit cavity conductance is given for different surface emissivities in several design handbooks (ASHRAE 1989). This conductance is the sum of the radiative and convective heat transfer coefficients. Here, we consider primarily the convective coefficient. Combined radiation and convection are considered in Section 2.4.

There are a number of relationships for heat transfer by convection across a rectangular cavity. These are usually correlations of experimental data in terms of three dimensionless parameters: the Nusselt number *Nu*, the Rayleigh number *Ra*, and the Prandtl number *Pr*.

The convective heat transfer coefficient (h_c) for vertical cavities (Figure 2.9) is conveniently determined as follows (see ElSherbiny *et al.* 1982):

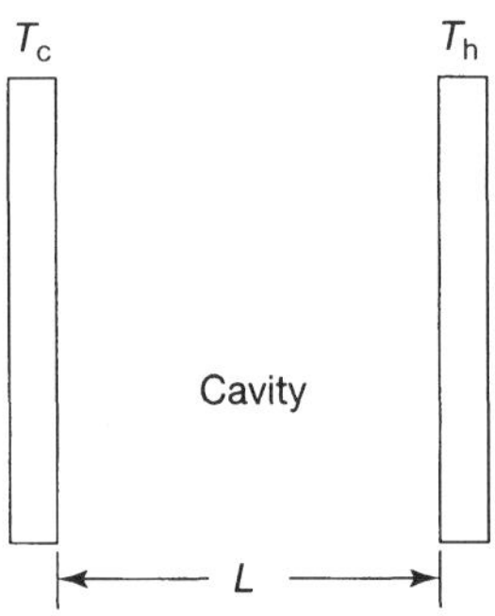

Figure 2.9 Cavity (window or in wall)

T_c = cold glazing temperature, T_h = hot glazing temperature
Thermal conductivity of air:

$$k_{air} = 0.002528 T_m^{1.5}/(T_m + 200) \quad (2.13a)$$

$$Ra = 2.737(1 + 2a)^2 a^4 (T_h - T_c) L^3 p^2 \quad (2.13b)$$

$$Nu_1 = 0.0605 Ra^{1/3} \quad (2.13c)$$

$$Nu_2 = [1 + 0.104 Ra^{0.293}/(1 + (6310/Ra)^{1.36})^3]^{1/3} \quad (2.13d)$$

$$h_c = \max(Nu_1, Nu_2) k_{air}/L \quad (2.13e)$$

The variation of heat transfer coefficient with gap thickness is shown in Figure 2.10. Note that the convective heat transfer coefficient h_c reaches its minimum value for a cavity width equal to 13mm.

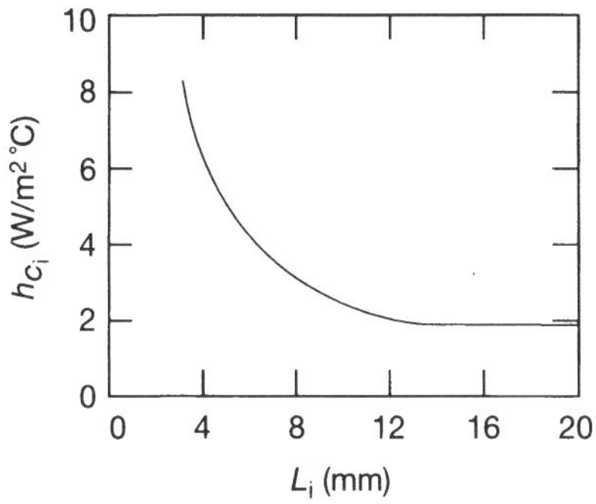

Figure 2.10 Variation of convective coefficient for a vertical cavity with width L_i

2.2.2 Convective heat transfer coefficients in rooms

For heat flow downward, that is conduction across the air-film, the following correlation is recommended (McAdams 1959). We apply it here to a cold floor. Laminar flow is assumed with Rayleigh number in the range 3×10^5 to 3×10^{10}.

$$h_c = 0.59[(T_s - T_{ai})/L]^{1/4} \tag{2.14}$$

where T_s = surface temperature, T_{ai} = air temperature and L = characteristic dimension (m).

Figure 2.11 shows the variation of h_c with T_s. Typically, an h_c in the range 0.6–0.7°C would be expected for cold floors in the winter.

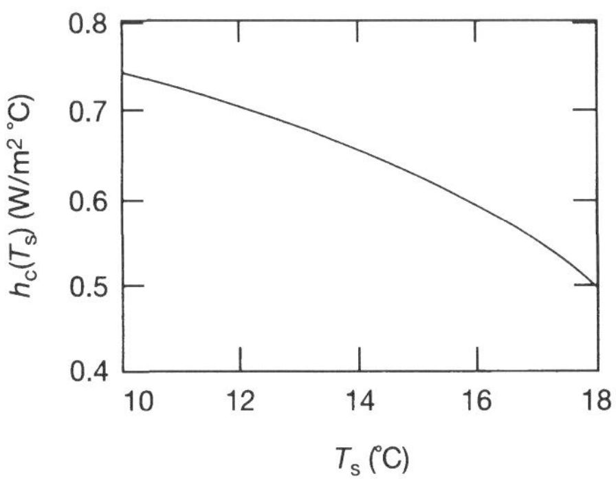

Figure 2.11 Variation of convective coefficient for heat flow downward – horizontal surface, T_{ai} = 20°C, L = 1 m

For **heat flow upward**, such as on a heated floor, the following turbulent flow relationship for air is recommended:

$$h_c = 1.52(T_s - T_{ai})^{1/3} \tag{2.15}$$

Figure 2.12 depicts the variation of h_c with surface temperature. Thus, for a floor heating system, if the average surface temperature is 26°C and the room temperature 20°C, then h_c is about 3.1W/m²°C.

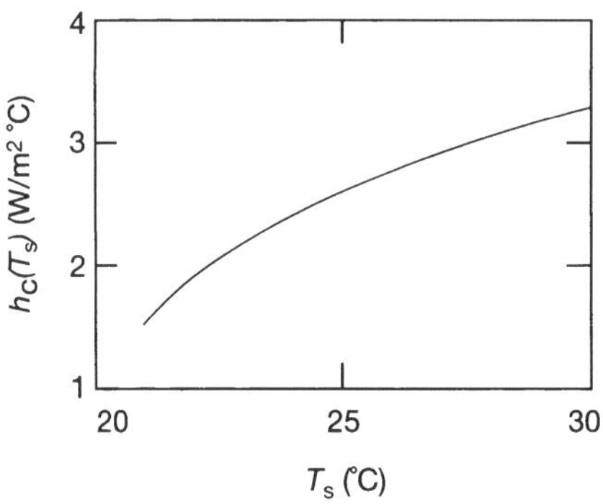

Figure 2.12 Variation of convective coefficient for heat flow upward – horizontal surface, T_{ai} = 20°C

For **vertical surfaces** we have horizontal heat flow, which is independent of direction. Because flow is well developed after a short distance from the bottom or top of a wall, it is generally turbulent, and the following correlation is often used for h_c (MacAdams 1959):

$$h_c = 1.31(T_s - T_{ai})^{1/3} \tag{2.16}$$

The Rayleigh number is assumed to be over 10^9.

The detailed variation of h_c with temperature difference is plotted in Figure 2.13. Generally, h_c would be expected to vary in the range 2–3 W/m²°C, most of the time being near the low end.

Note that the above relationships give values of h_c averaged over a wall – there could be considerable variation of the local h_c depending on flow development and local geometry. There is also considerable disagreement among different researchers and organizations on recommended values for h_c to be used in building energy analysis. Table 2.1 gives different values of h_c recommended by various organizations.

In residential buildings the air in each room is usually assumed to be at a uniform temperature. General equations, derived by various workers from correlations of

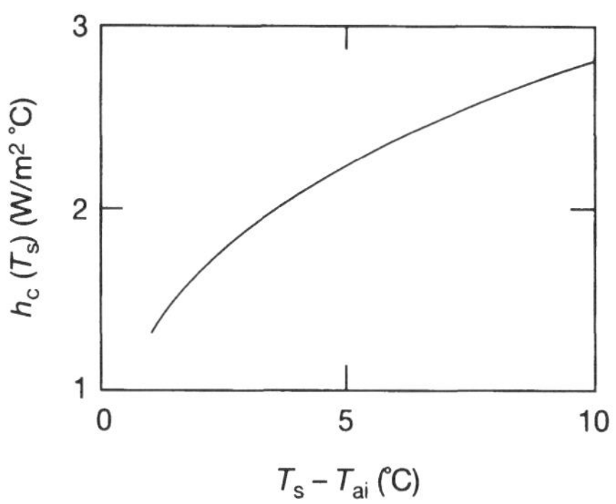

Figure 2.13 Variation of convective coefficient for heat flow at a vertical surface as a function of temperature difference

experimental results for convective heat transfer from plane surfaces, are described by Raithby and Hollands (1984). The heat flow depends on the surface inclination and whether the flow is laminar or turbulent. Equations for convection in buildings are given by ASHRAE (1989).

Convective heat transfer coefficients are typically linearized by calculating constant values based on assumed temperature differences between air and a surface. For routine analysis, detailed hourly simulation programs usually employ constant convective coefficients. The convective heat transfer coefficients suggested by ASHRAE, the corresponding British organization CIBSE, and by Kimura (1977), are given in Table 2.1

Table 2.1 Convective heat transfer coefficients (W/K/m^2) for winter conditions and still air

Surface	*Reference* *ASHRAE (1981)*	*CIBSE (1976)*	*Kimura (1977)*
Vertical	3.08	3.0	3.5
Horizontal – heat flow down	1.0	1.5	1.0
Horizontal	4.0	4.3	4.0

2.2.3 Natural convection between rooms

In passive solar buildings, where a significant temperature swing is permitted in a direct-gain room connected through a doorway to a north-facing room, temperature differences of 1–7°C may exist between the air in the two rooms, leading to a significant amount of heat being transferred by natural convection.

Bauman *et al.* (1983) developed a correlation based on the temperature difference between a hot driving surface in a room and a cold heated surface in the other room.

The interzone convective coefficient is given by

$$h_{iz} = 73C_{iz}(H_d T_{ab})^{1/2} \tag{2.17}$$

where H_d is the door height, and C_{iz} is a constant depending on room geometry (between 0.65 and 1.0). The temperature difference T_{ab} is equal to (T(1a) – T(1b)) for the case of two zones with temperatures T(1a) and T(1b); a constant value must be assumed for linearization (this is the standard procedure also used for surface convective coefficients).

2.3 Radiation heat transfer

Thermal radiation is electromagnetic radiation emitted by a body as a function of its temperature. All types of radiation are propagated at the speed of light, 3×10^8 m/s. The speed c is equal to

$$c = \lambda \nu \tag{2.18}$$

where λ = wavelength and ν = frequency.

Thermal radiation is in the range from about 0.1 to 100μm. The visible portion lies in the range of about 0.35–0.75μm. In building thermal analysis we generally separate thermal radiation into two regions – the short-wave region with wavelength less than or equal to 3μm and the long-wave region with wavelength greater than 3μm. More than 99% of the radiation emitted by the sun is short wave, while more than 99% of the radiation emitted by bodies at earth temperatures (–40°C to 45°C) is long wave.

Radiation may be also described with quantum mechanics, which stipulates that it is propagated as discrete quanta, each quantum having energy

$$E = h\nu \tag{2.19a}$$

where h = Planck's constant = 6.625×10^{-34} J s.

Based on the principle of relativity we also have

$$E = mc^2 \tag{2.19b}$$

Applying quantum-statistical thermodynamics, an expression for the radiation density may then be obtained per unit volume and unit wavelength as:

$$u_\lambda = \frac{8\pi h_c \lambda^{-5}}{\rho^{hc/(\lambda kT)} - 1} \tag{2.19c}$$

where k = 1.38066×10^{-23} J/(molecule K) (Boltzmann constant).

When the above expression is integrated over all wavelengths, we obtain the total thermal radiation emitted by a blackbody. Its emissive power is given by the Stephan–Boltzmann law:

$$E_b = \sigma T^4 \tag{2.20}$$

where $\sigma = 5.669 \times 10^{-8}$ W/m^2 K^4 and temperature is in kelvin.

The spectral emissive power of a blackbody is given by

$$E_{b\lambda} = C_1\lambda^{-5}/[\exp(C_2/\lambda T) - 1] \tag{2.21}$$

where λ = wavelength (μm), T = temperature (K), $C_1 = 3.743 \times 10^8$ W μm^4/m^2 and $C_2 = 1.4387 \times 10^4$ W μm/K.

The spectral emissive power is related to u_λ by

$$E_{b\lambda} = u_\lambda c/4 \tag{2.22}$$

A blackbody is a body that emits the maximum possible amount of radiation and absorbs all radiation incident on it – hence the term blackbody as it appears black to the eye.

Figure 2.14 shows the spectral emissive power of the sun, approximated as a black body at 6000K. The peak of all spectral emissive power curves may be shown (by differentiating equation 2.21 and setting it equal to 0) to satisfy Wien's displacement law:

$$\lambda_{max} T = 2897.6 \text{ μm K} \tag{2.23}$$

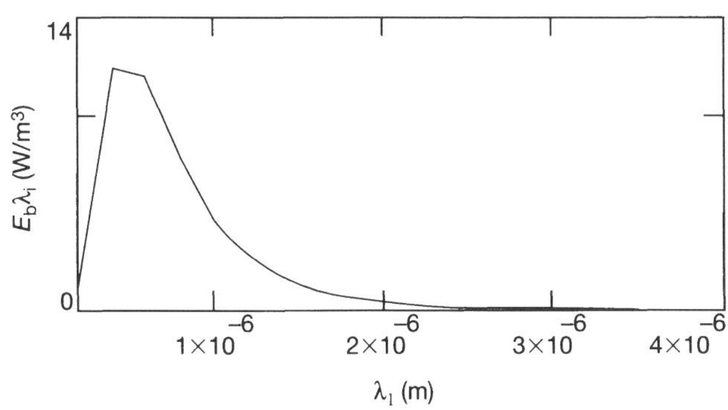

Figure 14 Spectral emissive power of the sun

Note that the peak emissive power of the sun lies in the visible region. The area under the curve is equal to the emissive power E_b given by equation 2.20.

We often need to determine the fraction of emitted radiation f_{12} in a given wavelength range λ_1–λ_2.

$$f_{12} = \frac{\displaystyle\int_{\lambda_1}^{\lambda_2} \frac{C_1}{\lambda^5\left(\exp\left(\frac{C_2}{\lambda T}\right)-1\right)} d\lambda}{\sigma T^4} \tag{2.24}$$

Example

A tungsten filament is heated to 2400 K. What fraction of the radiation is in the visible range?

Equation 2.24 is applied and evaluated numerically with $\lambda_1 = 0.38\ \mu\text{m}$, $\lambda_2 = 0.78\ \mu\text{m}$ and $T = 2400$ K.

$$f_{12} = 0.048$$

Therefore only 5% of emitted radiation is in the visible range.

RADIATION PROPERTIES

The absorptance of a surface is the fraction of total irradiation (incident radiation) absorbed by the body. The reflectance of the surface is equal to the fraction of the irradiation reflected from the surface. The transmittance of surface is the fraction of the irradiation transmitted by the surface (equal to 0 for opaque surfaces). An energy balance at the surfaces shows that

$$\rho + \tau + \alpha = 1$$

where ρ = reflectivity, τ = transmittance and α = absorptance.

Emissivity, absorptance, transmittance and reflectance generally vary with wavelength. It can be shown that the monochromatic emissivity and absorptance of a surface are equal.

$$\varepsilon_\lambda = E_\lambda / E_{b\lambda} \tag{2.25}$$

The total hemispherical emissivity of a surface is equal to the ratio of the total hemispherical emitted radiation to the total blackbody emissive power:

$$\varepsilon = \frac{\displaystyle\int_0^\infty \varepsilon_\lambda E_{b\lambda} d\lambda}{\sigma T^4} \tag{2.26}$$

Example

The total hemispherical emissivity of aluminium paint is 0.30 at wavelengths below 3 mm and 0.7 at longer wavelengths. Determine the total emissivity at a room temperature of 20°C:

$$\varepsilon = \frac{\int_{0.1\mu m}^{3\mu m} \varepsilon_s \frac{c_1}{\lambda^5\left[\exp\left(\frac{c_2}{\lambda T_1}\right)-1\right]} d\lambda + \int_{3\mu m}^{100\mu m} \varepsilon_l \frac{c_1}{\lambda^5\left[\exp\left(\frac{c_2}{\lambda T_1}\right)-1\right]} d\lambda}{\sigma T_1^4}$$

$\varepsilon = 0.697$

As can be seen, at low temperatures the emissivity is equal to the long-wave emissivity.

Grey bodies are surfaces with radiation properties whose values can be approximated as independent of wavelength. An acceptable approximation in building thermal analysis is selective grey surfaces; a surface may be assumed to be grey over specific wavelength ranges and associated temperature ranges.

For example, using the integration technique of the previous example, a long-wave emissivity of approximately 0.85–0.90 is determined for most building materials and paints (not metals) at ordinary temperatures. However, the solar (short-wave) absorptance of a glossy white surface, for example, can be as low as 0.2.

2.3.1 Calculation of view factors between room surfaces

The view factor, or shape factor F_{ij} from surface *i* to surface *j*, is equal to the fraction of diffuse radiation leaving surface *i* which is directly incident on surface *j*. The view factor between two surfaces at distance *r* is given by

$$A_1F_{1,2} = \int_{A_1}\int_{A_2}(\cos\varphi_1 \cos\varphi_2/\pi r^2)\mathrm{d}A_1\mathrm{d}A_2 \tag{2.27}$$

where *r* is the distance between the two areas and φ_1 and φ_2 are the angles between the local normals to the surfaces and the ray *r* connecting the two surfaces (Figure 2.15). The above equation may integrated analytically in some cases. When this is not possible, numerical integration techniques may be applied, replacing the integrals with summations over the discretized surfaces.

There are three main types of view factors between room surfaces:

- between surfaces at right angle,
- between parallel surfaces, and
- between the window and another surface.

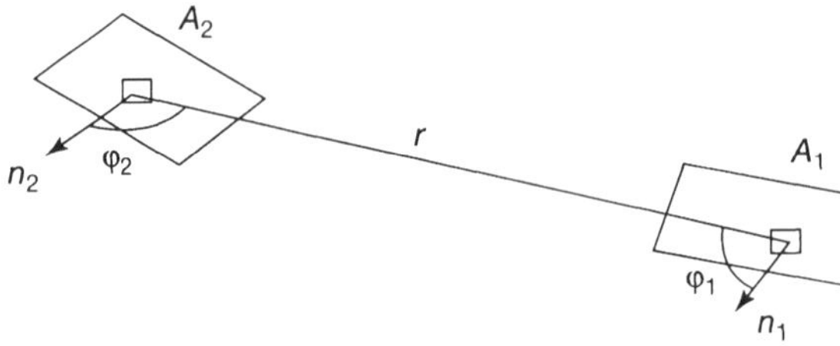

Figure 2.15 View factor F_{12} geometry

The view factors for the room below are determined after first calculating the view factor between two rectangular finite surfaces inclined at 90° to each other with one common surface as follows:

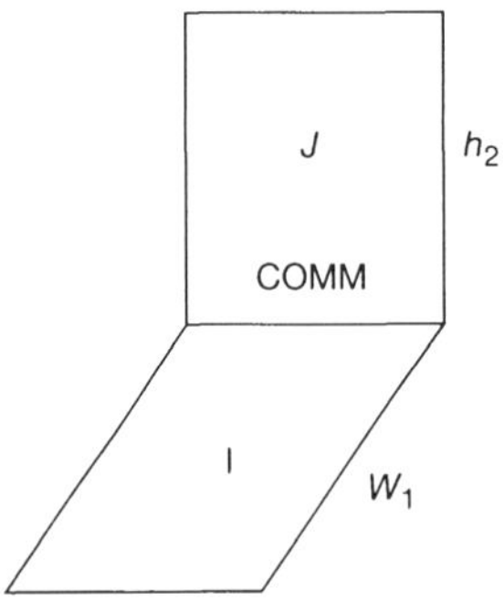

Define the following intermediate variables:

$$w = \frac{w_1}{comm} \qquad h = \frac{h_2}{comm}$$

$$A(h,w) = h^2 + w^2 \qquad B(w) = 1 + w^2 \qquad C(h) = 1 + h^2$$

$$D(h,w) = 1+(h^2 + w^2) \qquad E(w) = w^2 \qquad G(h) = h^2$$

View factor F_{ij} from *i* to *j*:

$$F_{ij}(w,h) = \frac{\left[w \cdot \operatorname{atan}\left(\frac{1}{w}\right) + h \cdot \operatorname{atan}\left(\frac{1}{h}\right)\right] - \sqrt{A(h,w)} \cdot \operatorname{atan}\left(\frac{1}{\sqrt{A(h,w)}}\right) + 0.25 \cdot \ln\left[\left(\frac{E(w) \cdot D(h,w)}{B(w) \cdot A(h,w)}\right)^{E(w)} \left(\frac{G(h) \cdot D(h,w)}{C(h) \cdot A(h,w)}\right)^{G(h)} \frac{B(w) \cdot C(h)}{D(h,w)}\right]}{\pi \cdot w}$$

Figure 2.16 Calculation of view factor between two perpendicular surfaces

The other view factors between the room surfaces are calculated by applying the following principles:

- Reciprocity

$$A_i F_{i,j} = A_j F_{j,i} \tag{2.28a}$$

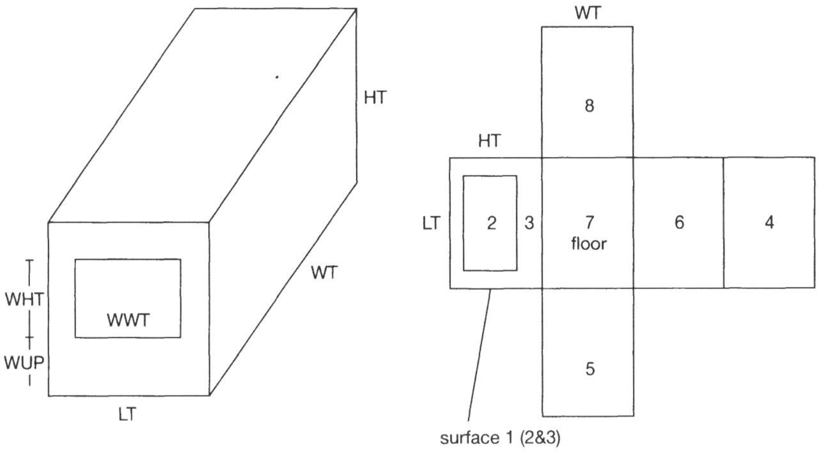

Figure 2.17 Room geometry for view factor calculation

- Symmetry:

  e.g. $F_{7,8} = F_{7,5}$

- Energy conservation:

$$\Sigma_j F_{i,j} = 1 \text{ for any surface } i \tag{2.28b}$$

Example

Calculate the view factors for a room with dimensions HT = 2.8 m, LT = 4.0 m, WT = 6.0 m, WWT = 3.0 m, WUP = 0.4 m, WHT = 1.6 m. We set COMM = LT in the function of Figure 2.16, W1 = HT and H2 = WT and we compute:

$$F_{6,7} = F_{ij}(\text{HT/LT, WT/LT}) = 0.268$$

Applying reciprocity:

$$F_{7,6} = A_6 F_{6,7}/A_7 = 0.125$$

From symmetry $F_{6,4} = F_{6,7}$ and $F_{4,6} = F_{7,6}$. Then for $F_{6,5}$ we set w1 = LT, h2 =WT and COMM = HT.

$$F_{6,5} = F_{ij}(\text{LT/HT, WT/HT}) = 0.191$$

Applying reciprocity we find $F_{6,5} = 0.127$.
From symmetry: $F_{8,5} = F_{5,6}$, $F_{1,5} = F_{6,8}$, $F_{5,1} = F_{8,6}$, $F_{1,8} = F_{6,8}$, $F_{8,1} = F_{8,6}$.

Using the function F_{ij} with w1 = HT, H2 = LT and COMM = WT we find:

$F_{8,4} = 0.276$

$F_{4,8} = 0.193$

From symmetry:

$F_{5,7} = F_{8,4}$

$F_{7,5} = F_{4,8}$

$F_{4,5} = F_{4,8}$

$F_{5,4} = F_{8,4}$

$F_{7,8} = F_{7,5}$

$F_{8,7} = F_{5,7}$

$F_{4,8} = F_{7,8}$

$F_{8,4} = F_{8,7}$

Now, we may apply equation 2.28b to determine the view factors between opposite parallel surfaces:

$F_{1,6} = 1 - 2 \times F_{6,8} - 2 \times \mathrm{F}_{6,4} = 0.082$

and $\mathrm{F}_{1,6} = \mathrm{F}_{6,1}$ (equal areas)

$F_{5,8} = 1 - 2 \times F_{5,4} - 2 \times F_{5,6} = 0.194$

and $F_{8,5} = F_{5,8}$

$F_{4,7} = 1 - 2 \times F_{4,8} - 2 \times F_{4,6} = 0.364$

and $F_{7,4} = F_{4,7}$

Now, using the same techniques and Figure 2.18 one may determine the view factors between the window and the other surfaces. Note that surface 1 is sub-divided into nine surfaces and so are the adjacent surfaces (we do not need to repeat the calculation for the other surfaces as the window is located in the centre of surface 1).

After similar calculations for Fab_c1c2d2 and Fa_c1c2, Fab_c2d, Fa_c2, Fb_d we may calculate the view factor between the window and the floor.

$$Fa\_2d := \frac{(Aab \cdot Fab\_c1c2d2 - Aa \cdot Fa\_c1c2 - Ab \cdot Fb\_2d)}{2 \cdot Aa}$$

$$Fa\_d := \frac{Aab \cdot Fab\_c2d - Aa \cdot Fa\_c2 - Ab \cdot Fb\_d}{2 \cdot Aa} \qquad F2\_a := (Fa\_2d - Fa\_d) \cdot \frac{Aa}{A2}$$

$$F_{2,7} := 2 \cdot F2\_a + F2\_b \qquad F_{2,7} = 0.27 \qquad F_{7,2} := A2 \cdot \frac{F_{2,7}}{A7} \qquad F_{7,2} := 0.054$$

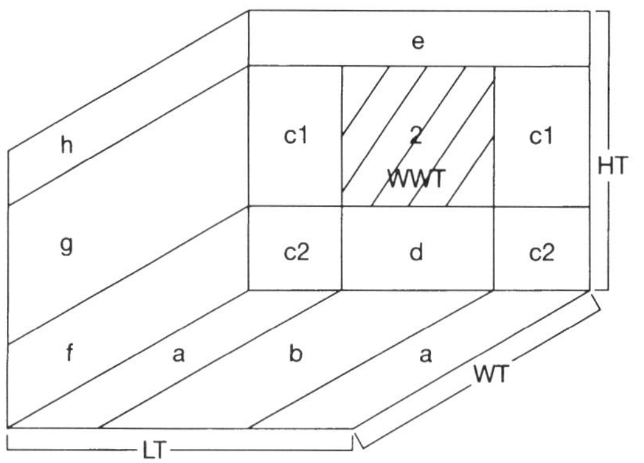

$Ab := WWT \cdot WT$

$Ad := WWT \cdot WUP$

$A2 := WWT \cdot WHT$

$DIS := \frac{LT - WWT}{2}$

$Ac1 := WHT \cdot DIS \qquad Ac2 := WUP \cdot DIS \qquad Aa := DIS \cdot WT \qquad Aab := WT \cdot (DIS + WWT)$

$w1 := WT \qquad h2 := WHT + WUP \qquad comm := WWT \qquad w := \frac{w1}{comm} \qquad h := \frac{h2}{comm}$

$Fb\_2d := Fij(w, h)$ ..F from Ab to A2 + Ad

Figure 2.18 Schematic used in calculation of view factors between window and walls

2.3.2 Radiative exchange between two surfaces

The net radiative heat exchange between black surfaces 1 and 2 emitting radiation E_{b1} and E_{b2} is given by

$$q_{r1-2} = A_1 F_{1,2}(E_{b1} - E_{b2}) = A_1 F_{1,2} \sigma (T_1^4 - T_2^4) \tag{2.29}$$

When grey surfaces exchange radiation we have a more complex relationship because part of the radiation incident on a surface is reflected, and in an enclosure we have an infinite number of reflections. In this case we may define a radiation exchange factor F' which is a function of view factors and surface emissivities.

$$q_{r1-2} = A_1 F'_{1,2}(E_{b1} - E_{b2}) = A_1 F'_{1,2} \sigma (T_1^4 - T_2^4) \tag{2.29a}$$

where $F'_{1,2} = f(F_{i,j}, \varepsilon_i)$.

The exchange factor between two parallel planes is given in the next section which calculates total thermal resistance in a cavity.

2.4 Convective and radiative heat transfer coefficients

2.4.1 Thermal resistance of cavities

The great majority of thermal insulation materials owe their resistance to heat flow to a cellular structure with isolated air cavities. Thus, cavities are also employed in walls, windows and other building components to reduce heat transfer, while simultaneously they retard transmission of water vapour due to capillary action; in windows they also transmit solar radiation.

Cavities usually contain air. However, in newer, high-technology windows, they may also contain a vacuum or rare gases such as argon in order to reduce convection heat transfer. The unit conductance is given for different emissivities of the two cavity surfaces in several design handbooks (ASHRAE 1989). The total heat transfer coefficient of a cavity is equal to the sum of the radiative (h_c) and convective (h_r) heat transfer coefficients.

Convective heat transfer through a cavity may be calculated by a number of correlations reviewed by Daskalaki *et al.* (1994). The variation of convective coefficient based on a correlation (ElSherbiny *et al.*) (Figs 2.10 and 2.11) for a vertical cavity as a function of width is given in Table 2.2. Values given are based on an outside temperature of 35°C and an inside temperature of 25°C.

Table 2.2 Convective heat transfer coefficient for sealed vertical cavities (summer conditions)

Cavity width (mm)	h_c *(W/m²°C)*
6	4.4
8	3.1
10	2.6
12	2.2
14	1.9

The cavity convective heat transfer coefficient does not change significantly for widths larger than about 13 mm. The change is even smaller for winter conditions. The radiative heat transfer coefficient between two parallel surfaces in a cavity is given by

$$h_r = \sigma F'_{12}(4T_m^3) \tag{2.30}$$

where

$\sigma = 5.67 \times 10^{-8}$ W/m^2 K^4 (Stefan–Boltzmann constant)

$T_m = (T_1 + T_2)/2$ (mean temperature, K)

$F'_{12} = 1/[(1/\varepsilon_1) + (1/\varepsilon_2) - 1]$ (radiation exchange factor)

The term $(4T_m^3)$ is a linearization factor for radiation heat transfer (approximately equal to $[T_1^4 - T_2^4]/[T_1 - T_2]$).

Most building materials have a high emissivity, about 0.85–0.95, and radiation usually accounts for about two-thirds of the heat transfer in a cavity. The radiative coefficient is about 5.0 W/m^2 K for regular glass surfaces with emissivities of about 0.9. When a low-emissivity coating is applied on one of the two surfaces facing the cavity h_r is reduced to 0.6 W/m^2 K, i.e. a dramatic reduction of about 90%. Therefore, low-emissivity coating (emissivity 0.05–0.1) is one of the most cost-effective techniques to reduce radiation heat transfer through a double-glazed window.

For a wall with an aluminium foil on one side and a width of 12 mm, the total heat transfer coefficient is therefore given by

$$h = h_c + h_r = 2.2 + 0.6 = 2.8 \text{ W/m}^2 \text{ K}$$

For a window, the value would be 7.4 without a low-e coating and 2.8 with a low-e coating. Note that this value is accurate near the centre of the window. Near the frame we have edge effects.

2.5 Bibliography

ASHRAE (1989) *ASHRAE Handbook – Fundamentals*, Atlanta, Georgia, USA.

ASTM (1989) Guarded Hot Box Test (C236) *ASTM Book of Standards*, ASTM USA.

ASTM (1989) Calibrated Hot Box Test (C976) *ASTM Book of Standards*, ASTM USA.

Athienitis A.K. (1998) *Building Thermal Analysis*, Electronic Mathcad Book, 2nd edn MathSoft, Boston, USA.

Bauman F., Anderson B. and Carroll W.L. (1981) 'Verification of BLAST by comparison with measurements of a solar-dominated test cell and a thermally massive building', *Proc. of ASME Solar Ener. Div. 3rd Annual Conf.*, Reno, Nevada, pp. 299–307.

Bauman F., Gadgil A., Kammerud R., Altmayer E. and Nansteel M. (1983) 'Convective heat transfer in buildings: recent research results', *ASHRAE Transactions,* **89** Pt.1, paper no. 2750.

Daskalaki E., Santamouris M., Balaras C.A. and Asimakopoulos D. (1994) 'Natural convection heat transfer coefficients from vertical and horizontal surfaces for building applications', *Energy and Buildings*, **20**, 243–249.

Dick H.V. and Linden G.P. (1994) 'PASSYS test method for thermal and solar characteristics of building components', European Conf. on Energy Performance and Indoor Climate in Buildings, Lyon, France, Nov., pp. 1092–1097.

Dijk H.A.L. and Linden G.P. (1994) 'The PASSYS test method for thermal and solar characterization of building components', European Conf. on Energy Performance and Indoor Climate in Buildings, Lyon, France, Nov., pp. 1092–1097.

ElSherbiny S.M., Raithby G.D. and Hollands K.G.T. (1982) 'Heat transfer by natural convection across vertical and inclined air layers', *Journal of Heat Transfer*, **104**:1, 96–102.

Energy in Architecture – The European Passive Solar Handbook (1992) EC Publication No. 13446, Batsford, London.

Goulding J.R., Lewis O.J., Steemers T.C. *Energy Conscious Design*, Batsford for CEC, London.

Harazono Y., Shigeaki T., Nakase I. and Ikeda H. (1990) 'Effects of rooftop vegetation using artificial substrates on the urban climate and the thermal load of buildings', *Energy and Buildings*, **15**:3–4, 435–442.

LBL (1988) Window 3.1 Program, LBL25-686, USA.

McAdams W.H. (1954) *Heat Transmission*, 3rd edn McGraw-Hill, New York.

Norlen U. and Bloem J.J. (1994) 'Comparison of methods for thermal resistance estimation', European Conf. on Energy Performance and Indoor Climate in Buildings, Lyon, France, Nov., pp. 804–880.

PASSYS II (1994) Final Report, CEC.

Raithby G. and Hollands K.G.T. (1984) 'Natural convection', in *Heat Transfer Handbook*, McGraw-Hill Inc., New York.

Santamouris M. (1994) PASSPORT Manual, CEC.

Sullivan H.F. (1987) VISION computer program, Univ. of Waterloo, Canada.

CHAPTER 3

Fenestration components and systems

3.1 Window thermal and solar properties

Fenestration is an essential component of any building enclosure, in the form of windows, glass doors, skylights, sunspaces or atria. It provides the path for visual and psychological communication with the external environment. Sunlight is an essential requirement for our everyday life.

Design of fenestration systems should consider all factors that affect performance and indoor environment:

- **Net heat transfer** across the fenestration system by conduction, convection and long-wave radiation; this is usually assessed as proportional to the window effective thermal conductance (*U*), measured with techniques such as the guarded hot box method (ASTM 1989).
- **Net transmitted solar radiation**, some of which enters the living space directly (as diffuse or beam solar radiation). A smaller portion of the incident solar radiation is absorbed by the fenestration and part of it is transferred to the interior of the zone as infrared radiation or by convection; ASHRAE (1989) uses the concept of solar heat gain factor (SHGF) to assess this. The PASSYS (1994) test procedures in Europe denote this as the total solar heat gain factor (*g*); *g* is defined as equal to the heat flux through the component under steady conditions for zero temperature difference between indoor and outdoor environment, divided by intensity of solar radiation incident on the component.
- **Daylighting effectiveness**: assessed as a function of the illumination of the working or living space at the desired locations, and a glare index. The total amount of solar radiation incident on each room surface also depends on the solar reflectance of the room interior surfaces. The daylight factor (DF) is a commonly used parameter to assess daylighting effectiveness; it is the ratio of illumination at the point of interest (e.g. work plane) to illumination outside the fenestration due to unobstructed sky.
- **Influence of fenestration interior temperature on thermal comfort**: if this temperature is too high or too low and the window area is large, it raises or

reduces the mean radiant temperature significantly, particularly near the window.

- **Acoustic performance**, noise transmission.

Aesthetic factors such as **external view** also depend on fenestration shape, landscape and surrounding vegetation, but are difficult to assess objectively based on scientific methods.

3.1.1 Transmission of solar radiation

The solar transmittance, reflectance and absorptance of windows or other transparent building components such as solar collectors need to be determined in order to calculate how much solar radiation they transmit.

First we determine the solar properties of a layer of glass with thickness L, refractive index n and extinction coefficient k. Consider Figure 3.1.

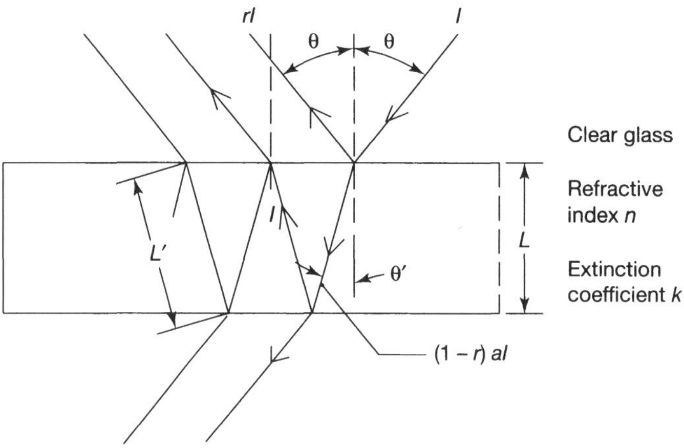

Figure 3.1 Transmission of solar radiation through glazing

Snell's law: $\sin \theta' = (\sin \theta)/n$ (3.1)

Component reflectivity (r) is the fraction of each ray component reflected. It is determined from the Fresnel relations of electromagnetic theory:

$$r = \frac{1}{2}\left[\left(\frac{\sin(\theta - \theta')}{\sin(\theta + \theta')}\right)^2 + \left(\frac{\tan(\theta - \theta')}{\tan(\theta + \theta')}\right)^2\right] \qquad L' = \frac{L}{\sqrt{1 - \left(\frac{\sin(\theta)}{n}\right)^2}} \tag{3.2}$$

The transmittance of transparent glazing decays exponentially with its extinction coefficient × thickness product. Some extinction coefficients are given in Table 3.1.

Table 3.1 Extinction coefficients

Type of glass	*Extinction coefficient k (per metre)*
Double strength, A quality	7.76
Clear plate glass	6.96
Heat-absorbing glass	132

The intensity $I(x)$ of radiation after it has travelled a distance x in a material is an exponential decay function of its intensity at $x = 0$, I_0 and the extinction coefficient k:

$$I(x) = I_0 e^{-kx} \tag{3.3}$$

Therefore, after travelling a distance L' as shown in Figure 3.1, the intensity is equal to

$$I(L') = I_0 a \text{ where } a = e^{-kL'}$$

The quantity $a = e^{-kL'}$ is the fraction of radiation available after each reflection in the glazing. Through simple ray-tracing techniques, one may show that

$$\tau = (1 - r)^2 a + r^2(1 - r)^2 a^3 + r^4(1 - r)^2 a^5 + \ldots$$

This is a convergent geometric series. Therefore, the effective transmittance of the glazing is determined from the limit of a geometric series as:

$$\tau = (1 - r)^2 a/(1 - r^2 a^2) \tag{3.4}$$

Similarly, the effective reflectance of the glazing is

$$\rho = r + [r(1 - r)^2 a^2/(1 - r^2 a^2)] \tag{3.5}$$

The absorptance is then determined from the energy conservation relationship which states that the incident energy is equal to the amount transmitted plus the amount reflected plus the portion absorbed:

$$\alpha = 1 - \rho - \tau \tag{3.5a}$$

Note that the above properties are a function of incidence angle and wavelength.

Typically properties are averaged over two wavelength ranges:

- solar: 0.3–3 μm
- long wave: 3–30 μm.

The averaged properties are determined as a weighted average of spectral values. For example, the solar absorptance is given by:

$$\alpha_s = \frac{\int_{0.3\mu m}^{3\mu m} \alpha_\lambda I_\lambda \, d\lambda}{\int_{0.3\mu m}^{3\mu m} I_\lambda \, d\lambda} \tag{3.6}$$

For efficient conversion of solar energy to heat, a high solar absorptance and a low long-wave emissivity are required. Selective surface coatings may be used to achieve such properties.

3.1.2 Double-glazed windows

For a double-glazed window, the effective transmittance t_e is usually required given τ_i, τ_o, ρ_2 and ρ_3 for the two glazings. The effective transmittance is given by:

$$\tau_e = \tau_i\tau_o/(1 - \rho_2\rho_3) \tag{3.7}$$

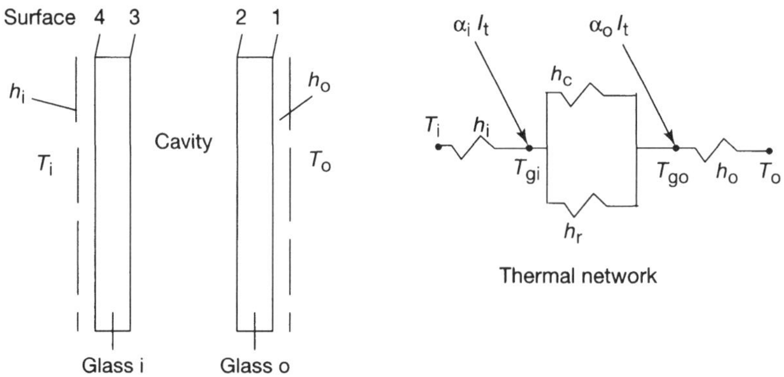

Figure 3.2 Double glazing and thermal network

In heat transmission calculations, the effective absorptance of each glazing is also required:

$$\alpha_o = \alpha_1 + \alpha_2\tau_0\rho_3(1 - \rho_2\rho_3) \tag{3.8}$$

$$\alpha_i = \alpha_3\tau_0(1 - \rho_2\rho_3) \tag{3.9}$$

where
α_o = absorptance of outer glass
α_i = absorptance of inner glass
α_2 = absorptance of outer glass for solar radiation incident on indoor surface (2)
α_3 = absorptance of glass i for solar radiation incident on surface (3)
ρ_j = reflectance of surface *j*

Total solar radiation aborbed in the outer glass is $I_{ao} = \alpha_o I_t$.
Total solar radiation aborbed in the inner glass is $I_{ai} = \alpha_i I_t$.

Note that for uncoated surfaces the directional reflectance (and absorptance) of the glazing surfaces are equal. (Note: the diffuse component of *I* should be multiplied by the diffuse absorptance for a more accurate calculation.)

Part of the solar radiation absorbed in the glazings is transmitted by convection or as long-wave radiation heat into the room. An energy balance at the two glazings will show that the total amount of solar radiation that flows as heat into the room (after absorption in the glazings) is:

$$q_{sa} = \frac{U}{h_o}(\alpha_o I_t) + U\left(\frac{1}{h_o} + \frac{1}{h_g}\right)(\alpha_i I_t) \tag{3.10}$$

where

$$U = \frac{1}{\left(\frac{1}{h_o} + \frac{1}{h_g} + \frac{1}{h_i}\right)} \quad \text{(window U – value)} \tag{3.10a}$$

h_o = outside film coefficient
h_g = cavity heat transfer coefficient = $h_c + h_r$
h_i = inside film coefficient

The variation of normal transmittance and absorptance (for incidence angle 0) is given in Figure 3.3 for various *kL* (extinction coefficient × thickness value) values. As can be seen, for clear glass 5 mm thick (*kL* is equal approximately to 0.04) the transmittance is 0.86 while for *kL* = 0.65, which is heat-absorbing glass, the transmittance is 0.47. In the latter case, the solar absorptance is also about 0.47.

As can be seen from Figure 3.4, the transmittance of glazing for direct radiation remains relatively steady until about 50° and drops fast after 60°. It is often assumed that the transmittance for diffuse radiation is equal to the beam transmittance at 60°.

Glazings may be specially treated to control their optical properties, e.g.:

- body-tinted glass with high absorptance
- surface-coated glasses with high directional reflectivity
- variable-transmission glasses

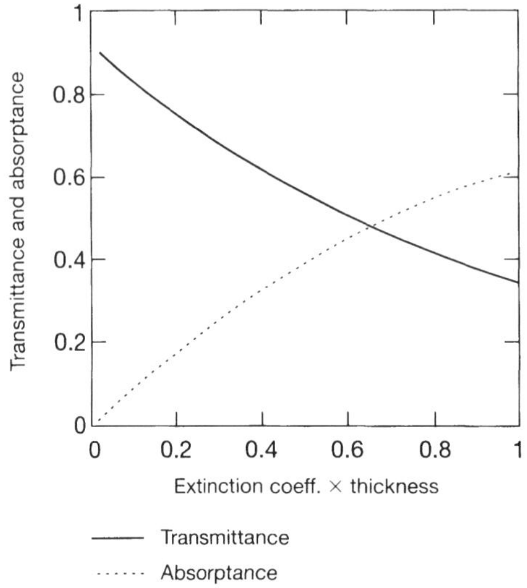

Figure 3.3 Transmittance and absorptance for various kL values

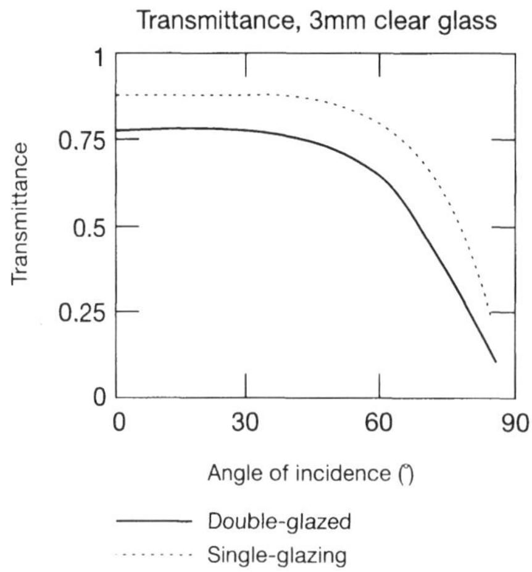

Figure 3.4 Transmittance of single and double glazing

- translucent glazings which diffuse beam solar radiation
- special sun-control membranes
- electrochromic coatings whose properties may be electrically controlled.

Body-tinted (absorptive) glasses have high extinction coefficient, low transmittance and high absorptance. The low transmittance also reduces the amount of daylight transmitted. The body-tint is produced by adding small quantities of metal oxides (iron, cobalt, selenium) which usually also give an apparent colour to the glass – green, blue, bronze or grey.

Surface coatings applied on one or both surfaces may modify its long-wave or short-wave radiation properties. They are usually reflective or low-e coatings.

Reflective glasses are used for solar control by increasing the reflection of solar radiation with special coatings of silver, gold, copper, etc.; these glasses usually have high absorptance.

Low-emissivity (low-e) glass has a special coating on one surface which reduces its long-wave (for wavelengths greater than 3 μm) emissivity from about 0.9 for regular glass to about 0.1. The radiative heat transfer coefficient for the cavity in a double-glazed window is then dramatically reduced as described above. Special coatings that allow high solar gains into a room while minimizing conduction heat losses through the glazing are very effective in passive solar heating. They are also useful for cooling in that they reduce the window *U*-value and summer heat gains, provided they are properly shaded. Typical coatings consist of three layers – a thin metal layer (usually gold, silver or copper) sandwiched between dielectric layers of tin oxide.

Variable-transmission glasses are a new development that permit the building envelope to be used dynamically, responding to outdoor climate and interior thermal needs. The transmittance may be varied by photochromic, thermochromic and electrochromic films.

The principle of operation of **electrochromic glazing** (Cogan *et al.* 1986) is based on the change of the optical properties of certain laminated materials when subjected to an external electric field. This type of glazing has an electrochromic coating. The coating is activated by a small electric voltage, generated by the building automation system or manually by the occupants. This voltage changes the tint of the coating and therefore its transmittance. The original transmittance is restored by reversing the voltage. Currently there are only small-scale applications, such as rear-view mirrors in cars. These products vary their transmittance between 15% and 70%. Their life expectancy is presently rather short. This is a major drawback for their generalized use in buildings.

Thermochromic glazing (Lee *et al.* 1986) is made from tungsten trioxide or vanadium dioxide and can be applied on windows as a thin film. It passively switches between a heat-transmitting and heat-reflecting state. Therefore it can reduce the cooling loads and provide solar protection.

Translucent glazings have low solar transmittance and they diffuse daylight; they are suitable for atria and skylights where one does not need visual communication with the exterior environment.

3.2 Transmitted solar radiation and its distribution

3.2.1 Performance evaluation

Two important parameters commonly employed in comparing fenestration systems are the solar heat gain factor (SHGF) and the shading coefficient (SC) (ASHRAE 1989). These parameters are based on an energy balance of the system.

The total instantaneous rate of heat transfer through a glazing element may be determined from an energy balance for unit area of the fenestration and its thermal environment:

> Total energy transfer through window = heat transfer due to temperature difference between inside (T_i) and outside (T_o) + solar radiation transmitted through glazing + inward flow of solar radiation absorbed in glazing

For single glazing, the above energy balance may be written as follows:

$$q = U(T_o - T_i) + \tau_b I_b + \tau_d I_d + (U/h_o)(\alpha I) \tag{3.11}$$

where the conductance U for single glazing is given by:

$$U = 1/[(1/h_i) + (1/h_o)]$$

$$\text{Total incident solar radiation} = I = I_b + I_d$$

The above equation may be written in the form

$$q = U(T_o - T_i) + (\tau_{eff} + U\alpha/h_o)I \tag{3.12}$$

where τ_{eff} is the effect transmittance for total solar radiation.

The solar heat gain (SHG) is the second term of equation 3.12, and may be expressed as

$$\text{SHG} = FI = (\tau_{eff} + U\alpha/h_o)I \tag{3.13}$$

where the dimensionless ratio of solar heat gain to incident solar radiation is given by

$$F = (\tau_{eff} + U\alpha/h_o)$$

The solar heat gains through sunlit double-strength (3 mm) sheet glass (DSA) are defined as the SHGF. Clear day SHGF are tabulated for various latitudes and dates by ASHRAE (1989). However, in polluted city centres SHGF may be up to

20% lower, for humid locations up to 30% lower, and for high elevations up to 15% higher than tabulated values.

The solar heat gain of any fenestration system, including curtains, etc. is related to the standard SHGF using a **shading coefficient** (SC). The SC is given by the ratio of solar heat gain of the fenestration to the solar heat gain of DSA glass:

$$\mathrm{SC} = \mathrm{SHG\ of\ fenestration/SHGF\ for\ DSA\ glass} \tag{3.14}$$

Therefore, SC may be written as

$$\mathrm{SC} = F \mathrm{\ of\ fenestration}/F \mathrm{\ of\ DSA}$$

$$\mathrm{SC} = F/0.87 \tag{3.15}$$

Thus

$$\mathrm{SC} \times \mathrm{SHGF} = (\tau_{\mathrm{eff}} + U\alpha/h_{\mathrm{o}})I \tag{3.16}$$

and equation (3.11) for total heat transfer through the glazing may be written as

$$q = (\mathrm{SC})(\mathrm{SHGF}) + U(T_{\mathrm{o}} - T_{\mathrm{i}}) \tag{3.17}$$

or as

$$q = FI + U(T_{\mathrm{o}} - T_{\mathrm{i}})$$

The most general parameters that are universally applicable are the thermal conductance (U-value) and the dimensionless F-factor of the fenestration. For double glazing (Figure 3.5) these parameters are given by

$$U = 1/[(1/h_{\mathrm{i}}) + (1/h_{\mathrm{g}}) + (1/h_{\mathrm{o}})] \tag{3.18a}$$

where h_{i} and h_{o} are the interior and exterior film coefficients respectively and h_{g} the cavity heat transfer coefficient.

$$F = \tau_{\mathrm{eff}} + U(\alpha_{\mathrm{o}}/h_{\mathrm{o}}) + [(U/h_{\mathrm{o}})+(U/h_{\mathrm{g}})]\alpha_{\mathrm{i}} \tag{3.18b}$$

where the first term is the effective transmittance, the second term is the factor for inward flow of solar radiation absorbed in the outer glazing (absorptance α_{o}) and the third term is the factor for inward flow of solar radiation absorbed in the inner glazing (absorptance α_{i}).

The overall thermal conductance U_{o} of a fenestration system depends on the

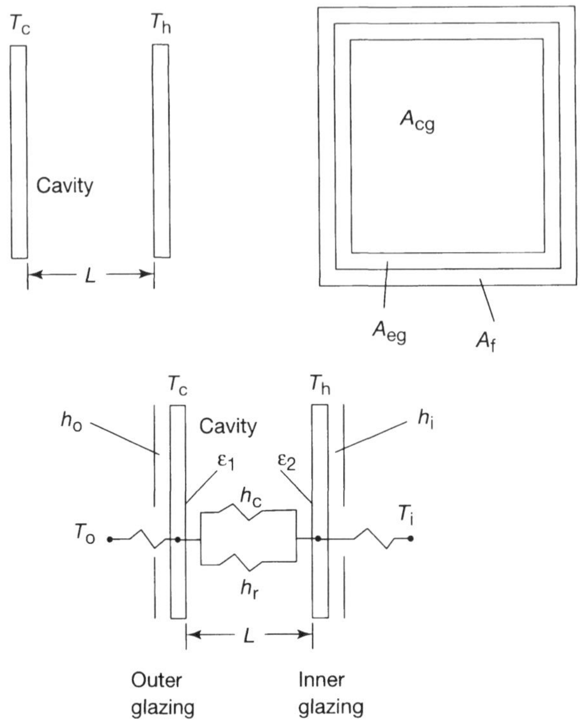

Figure 3.5 Double-glazing, area definitions and thermal network indicating heat flows

type of framing employed and the heat transfer regime at the edge of the glass. ASHRAE (1989) recommends subdivision of a window into three regions (see Figure 3.5) in calculating U_o: centre of glass region with area A_{cg} and conductance U_{cg}, edge of glass region with parameters A_{eg} and U_{eg}, and frame with parameters A_f and U_f. Since we have three heat flow paths in parallel U_o is given by:

$$U_o = (U_{cg}A_{cg} + U_{eg}A_{eg} + U_fA_f)/(A_{cg} + A_{eg} + A_f)$$

Centre-of-glass U-values are calculated based on one-dimensional heat flow correlations (Elsherbiny *et al.* 1982). Insulating glass units have continuous members around the perimeters to separate glazings and to provide an edge seal. Aluminium spacers are highly conductive and act as thermal bridges, greatly increasing heat transfer. The edge region is typically about 6 cm wide, adjacent to the frame. Aluminium frames sometimes have thermal breaks (wood or plastic inserts) that reduce U_f and U_{eg}. Vinyl frames have a much lower U-value. ASHRAE (1989) recommends the following values for frames:

Aluminium without thermal break: $U_f = 10.8\ \text{W/m}^2\ \text{K}$

Aluminium with thermal break: $U_f = 5.68\ \text{W/m}^2\ \text{K}$

Vinyl: $U_f = 2.27\ \text{W/m}^2\ \text{K}$

Note that the above values are approximate and may vary significantly for different company products owing to the three-dimensional character of the flow around window edges. The *ASHRAE Handbook* presents *U*-values for various windows including 1–3 glazings (glass, acrylic or polyester film), spacings of 6, 10 and 13 mm, emittances of 0.84, 0.4 and 0.15, and gas fills of air and argon.

Public domain programs such as WINDOW (LBL 1987) and VISION (Sullivan 1987) may also be used in computing U_o.

The transmission properties and shading coefficients for various types of window are shown in Table 3.2. It may be seen that thickness is not important for reflective glasses, although it is for body-tinted glasses.

Table 3.2 Transmission properties and shading coefficients for various types of glazing and window (Button and Rte 1992)

Glass type *Thickness*	*Light/heat ratio*	Light τ	ρ	Solar radiation α *(direct)*	τ *(total)*	Shading coefficient (SC)
Clear float						
4 mm		0.89	0.08	0.11	0.86	0.98
6 mm		0.87	0.08	0.15	0.83	0.95
10 mm		0.84	0.07	0.23	0.78	0.89
Reflective						
Silver						
6 mm	10/23	0.10	0.38	0.60	0.23	0.26
10 mm	10/23	0.10	0.37	0.60	0.23	0.27
6 mm	20/34	0.20	0.23	0.66	0.34	0.39
Bronze						
6 mm	10/24	0.10	0.19	0.73	0.24	0.27
10 mm	10/24	0.10	0.18	0.76	0.24	0.27
Blue						
6 mm	10/24	0.20	0.20	0.64	0.33	0.38
10 mm	10/24	0.20	0.20	0.66	0.33	0.38
Body-tinted						
Green						
6 mm	76/62	0.72	0.06	0.49	0.62	0.72
Blue						
6 mm	54/62	0.54	0.05	0.49	0.62	0.72
Bronze						
6 mm	50/62	0.50	0.05	0.36	0.70	0.72
10 mm	33/51	0.33	0.04	0.49	0.62	0.59

Example: Solar radiation transmitted through double glazing

We have seen how solar radiation incident on an inclined surface is calculated and how window transmittance is determined as a function of time. Now these methods are used together to calculate the instantaneous and daily total solar radiation transmitted through single-glazed and double-glazed windows. Consider a south-facing vertical window located in Montreal (latitude 45°N).

Determine the solar radiation transmitted for January 1 and June 30 for a double-glazed clear window.

Geometry: $L = 45°\text{N}, \beta = 90°, \psi = 90°$

For January 1

$n = 1$

For June 30

$n = 31 + 28 + 31 + 30 + 31 + 30 = 181$

Assume ground reflectance is 0.2. First perform calculations for 1 January:

$\delta = 23.45\sin[360(284 + n)/365)] = -23°$

The sunrise time is found and divided into eight equal time intervals.

$t_s = h_s/15 = \min\{\arccos[-\tan(L)\tan(\delta)], \arccos[-\tan(L - \beta)\tan(\delta)]\}/15$
$= 4.324 \text{ h}$

Time interval

$\Delta t = t_s/8 = 0.541 \text{ h}$

The position of the sun is then computed at the different times from noon to sunset (we only need to perform the analysis for half a day as the movement of the sun is symmetric about solar noon relative to a south-facing surface).

$h_j = t_j 15°/\text{h}$

where $t_j = j \times \Delta t$ with $j = 0, 1, 2, 3, 4$ ($j = 4$ corresponds to sunset).

$\sin \alpha_j = \cos L + \cos \delta \cos h_j + \sin L \sin \delta$ (solar altitude)

$\cos \varphi_j = (\sin \alpha_j \sin L - \sin \delta)/(\cos \alpha_j \cos L)$ (solar azimuth)

The incidence angle is then calculated:

$$\cos\,\theta_j = \cos\,\alpha_j\,\cos\,\gamma_j\,\sin\,\beta + \sin\,\alpha_j\,\cos\,\beta$$

where $\gamma = \varphi - \psi$.

The beam atmospheric transmitted is calculated (using Mathcad (MathSoft 2000)) based on Hottel's clear sky model:

$$A = 0.5\ \text{km}$$

The atmospheric transmittance for beam radiation is given by:

$$\tau_b = a_0 + a_1\,\exp[-k/\cos(z)] \tag{3.19}$$

where the constants a and k depend on climate and altitude A (km).

$$a_0 = 1.03[0.4327 - 0.008\ 21(6 - A)^2]$$

$$a_1 = 1.01[0.5055 - 0.005\ 95(6.5 - A)^2]$$

$$k = 1.0[0.2711 + 0.018\ 58(2.5 - A)^2]$$

After determining the beam transmittance of the atmosphere, we may also determine the diffuse transmittance (Liu and Jordan 1960):

$$\tau_d = 0.271 - 0.294\tau_b$$

The solar properties of a single glazing are then computed using the equations given in Section 3.1 (assuming $kL = 0.03$), and these are used in the calculation of the double-glazed window properties. The results are shown in Figure 3.6.

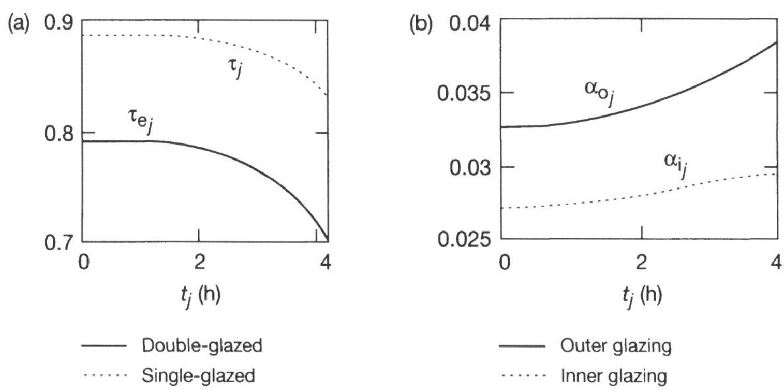

Figure 3.6 (a) Effective transmittance for double and single glazing; (b) absorptance of inner and outer panes

In calculating how much diffuse solar radiation is transmitted, it is assumed that the window diffuse transmittance is equal to the beam transmittance at an angle of incidence equal to 60°.

The total solar radiation I_t incident on the surface is given by the sum of the direct (beam) component I_b, the diffuse sky component I_{ds} and the diffuse solar radiation reflected from the ground I_{dg}.

$$I_t = I_b + I_{ds} + I_{dg}$$

where

$$I_b = I_{on}\tau_b \cos\theta$$

with the extraterrestrial solar radiation given by

$$I_{on} = 1353[1 + 0.033\cos(360n/365)]$$
$$I_{ds} = I_{on}\sin\alpha\ \tau_d(1 + \cos\beta)/2$$
$$I_{dg} = I_{on}\sin\alpha(\tau_b + \tau_d)\rho(1 - \cos\beta)/2$$

The transmitted solar radiation is then determined by multiplying the diffuse ($I_{ds} + I_{dg}$) radiation by the window diffuse transmittance and I_b by the window transmittance at the corresponding time:

$$G_t = \tau_e I_b + \tau_{ed}(I_{ds} + I_{dg})$$

Figure 3.7 shows the beam (G_{2b}) and diffuse (G_{2d}) parts of the transmitted radiation.

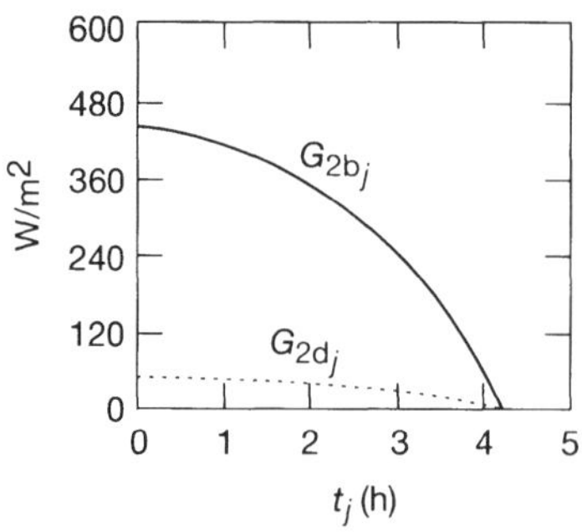

Figure 3.7 Transmitted beam and diffuse solar radiation for double-glazed clear window in January

The total transmitted daily solar radiation per square metre is twice the area under the curves in Figure 3.7, 10.54 MJ/m^2.

The solar radiation absorbed in each of the two glazings is given in Figure 3.8. The glazings absorb a small amount of radiation as they are clear (low extinction coefficient). Thicker glazing with body tint would absorb a significantly higher amount of solar radiation. In this case, the total solar radiation absorbed in the outer and inner glazings is 0.46 MJ/m^2 and 0.38 MJ/m^2 respectively. The portion transmitted inwards may be estimated by means of equation 3.10. The corresponding results for 30 June are given in Figure 3.9.

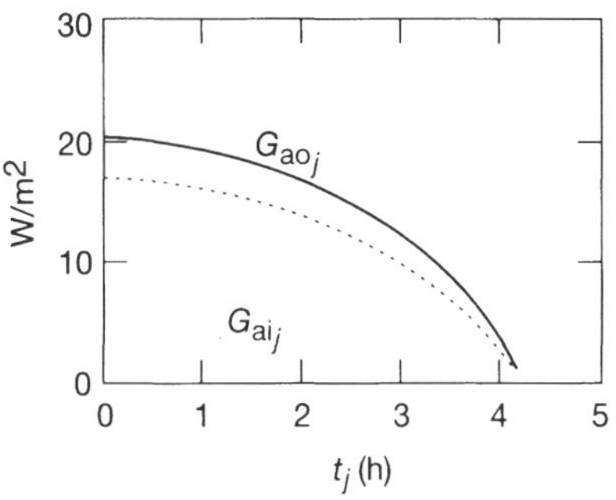

Figure 3.8 Solar radiation absorbed in inner (G_{ai}) and outer (G_{ao}) glazings

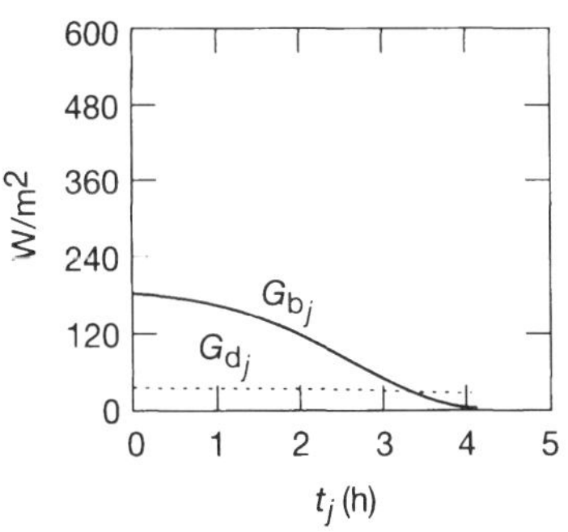

Figure 3.9 Transmitted beam and diffuse solar radiation for double-glazed clear window in June

Comparison of the results in Figures 3.7 and 3.9 reveals a very important result and useful principle for passive solar design: a south-facing window transmits a high amount of solar radiation in the winter and a low amount in the summer. This is due to the low incidence angles of the sun on the window in the winter, and the high incidence angle in the summer; at incidence angles in excess of 60° the window transmittance falls rapidly. In this case all incidence angles on 30 June are over 60°, whereas on 1 January they vary from a low of 22° at solar noon to 55° at sunset.

3.3 Transparent insulation

Transparent insulation materials (TIM) combine the properties of good optical transmission (described by the total solar transmittance) and good thermal insulation (described by the heat loss coefficient or *U*-value). One of the widely known applications of transparent insulation materials is in the cladding of building walls, replacing conventional opaque insulation.

Utilization of a TIM in a solar collector wall has been proved to have a significant potential to reduce heating load in passive solar buildings (Twidell *et al.* 1994). Transparent insulation converts one or more walls of the building into solar collectors. Initially, a single glazing was used as the transparent cover of a wall. This was known as a Trombe wall system (Trombe *et al.* 1977). Cellular transparent materials were introduced to improve the solar transmittance and thermal resistance of the wall system (Olsen 1982, Goetzberger *et al.* 1984) after being applied to solar collectors (Francia 1961, Hollands 1965). These materials enhance thermal resistance, in addition to having high solar transmittance. Other types of TIM are in use in the industry such as capillary glasses and aerogels. TIM classes have been reviewed by Platzer (1994) and they fall broadly into four categories:

- Type A: with material structures parallel to the surface plane (multiple glazing, plastic films, IR-reflective glass).
- Type B: with material structures perpendicular to the absorber (parallel slats, honeycombs, capillaries).
- Type C: scattering structures (duct plates, foams, and bubbles).
- Type D: quasi-homogeneous materials (glass fibres, aerogels).

The most popular type used today is type B, the honeycomb-structured materials made from plastics such as acrylic or polycarbonate (Platzer 1997). These materials have very high solar transmission with good insulation properties.

In a passive solar system, the thermal radiative heat losses from the absorber can be reduced by application of selective coatings. Once radiation losses have been restricted, natural convection dominates heat losses. Minimizing this convective heat loss requires the determination of some parameters such as (Hollands *et al.* 1978)

- the temperature difference between the absorber and the cover
- the distance between the absorber and the cover
- the angle of inclination of the air layer

- the air properties
- the geometry of the enclosure.

One of the methods for reducing the convective heat transfer is to divide the enclosure into a large number of cells (Figure 3.10). Owing to the reduced dimension of each cell, in comparison with the enclosure, the viscous forces acting on the air in each cell are increased and the movement of air in the cell is restricted. If the cell were sized correctly to maintain stagnant air (motionless air), the natural convection between the absorber and the cover would be suppressed. This gives an opportunity to increase the distance between the absorber and the cover and therefore to augment the insulating value of the air layer. Thus, honeycombing the air space will improve the thermal resistance of the wall system by means of convection suppression.

The honeycomb cell size must be chosen so as just to suppress the free convective heat transfer at the value of the design temperature difference. Use of smaller cell sizes will result in more of the expensive honeycomb being used than is actually necessary and will also cause unnecessary loss in solar transmission of the honeycomb. However, using a cell size larger than needed to suppress convection will result in greater heat loss. These factors underline the importance of properly choosing the cell size.

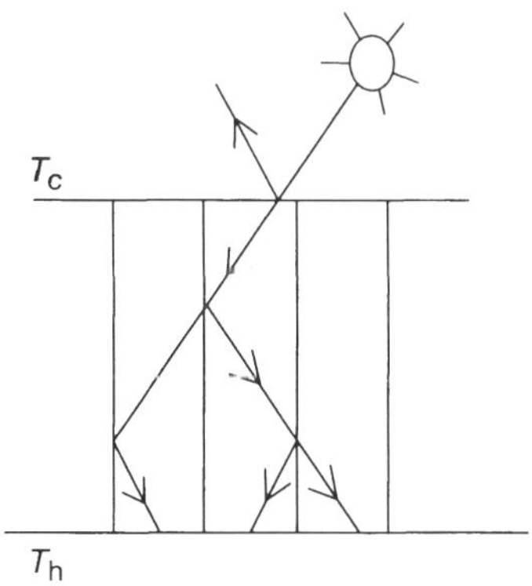

Figure 3.10 Trajectory of solar rays through honeycomb transparent insulation

The advantage of the honeycomb structure is that the solar radiation striking the insulation panel is reflected and re-reflected by the walls in the forward direction and thus reaches the absorber (Figure 3.10). Thus, if reflections are perfectly specular and there is no scattering or absorption in the walls, the solar transmission will be 100% (Hollands 1988). However, the transmittance of the honeycomb insulation is dependent upon the angle of incidence of solar radiation, i.e. it will be affected by the sun's position throughout the year.

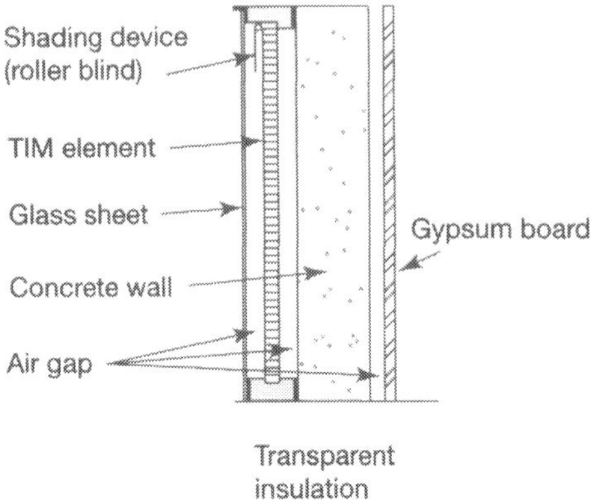

Figure 3.11 Transparently insulated wall (Athienitis and Ramadan 1999)

The high solar transmittance of the transparent honeycomb is matched with a good thermal insulation property. The conditions for having good insulation with low *U*-value (total heat transfer coefficient) may be described as follows:

- The structure must be thick enough to prevent heat conduction by air.
- The thickness of honeycomb walls must be low to minimize heat conduction through the transparent material.
- For reduction of radiative heat transfer, one practical solution is the use of a selective coating on the absorber. However, this should be paired with the introduction of an air gap between the TI unit and the absorber to minimize the coupled radiative and conductive heat transfer induced by the physical contact between TI and the absorber (Hollands *et al.* 1984).

A typical transparently insulated wall is shown in Figure 3.11. The transparently insulated wall system consists of an exterior single glazing protector, an air cavity, a transparent honeycomb material, a second air cavity, a concrete layer, a third air cavity and an interior gypsum board layer. The first air cavity permits the use of a shading device (roller blind) to enhance the thermal performance of the TI wall. The second cavity reduces the heat transfer between the absorber and the TI unit, especially when low-emissivity coatings cover the thermal mass (Hollands and Iyankaran 1985). The third air cavity increases the thermal resistance of the TI wall system and provides space for electrical wiring and other services. Heat transfer through air cavities is calculated based on the model of ElSherbiny *et al.* (1982) (see Chapter 2).

The plastic transparent honeycomb material is modelled as a square cell. The cross-sectional area of the cell is determined to suppress convective heat transfer which may occur between the protector and the absorber prior to the introduction of the TIM (Hollands *et al.* 1976). The aspect ratio of the cell, which is defined as the thickness of TIM over the hydraulic diameter of the cell, must be high to reduce the conductive (air conduction) and radiative heat transfer of the TIM layer (Tabor 1969). In addition, the thickness of the honeycomb wall material should be low to reduce the conductivity of the TIM material (about 50 μm).

Blinds have been introduced in the wall collector system to prevent overheating during days with excessive solar radiation (Twidell *et al.* 1994). The operation of the blind must satisfy the thermal comfort conditions. Thus, the blind is opened during winter days to allow the transmission of solar radiation to the thermal mass, and closed in winter nights to reduce heat losses from the heating space. In addition, closing the blind during cloudy winter days will contribute to the reduction of heating load.

Several researchers have examined the radiation heat transfer reduction and convection suppression properties of the transparent honeycomb materials. They focused first on the idea that, for large aspect ratio, the cell structure acts as a thermal radiation shield (Perrot 1967). However, as the reduction of thermal radiation heat transfer can be achieved by the selection of the absorber's material and coating, the convective heat transfer will dominate.

Honeycomb structures were used to reduce the convective heat losses between the cover and the absorber (Hollands 1965, Edwards 1976, Charters and Peterson 1972, Guthrie and Charters 1982). Generally, reduction of cell size reduces convection heat transfer. Hollands (1965) demonstrated that a cell size range between 0.4 and 0.6 inch (10–15 mm) suppresses convection for aspect ratios equal to 1 and 5 respectively. The use of glass has the advantage of better ultraviolet thermal stability than plastics. But the honeycomb wall thickness can be reduced with the use of plastics, resulting in less conduction heat loss and material conservation.

Using plastic honeycombs as an alternative to expensive glass partially solves the cost problem but creates other difficulties. If the wall covers a big area with a honeycomb cell, the cost of the wall material and manufacturing may become high. Furthermore, a thick wall of honeycomb cell may induce conduction. Therefore, the trend is towards very thin transparent materials (about 50 μm). However, care must be taken during transportation of these very thin materials. Also, their long-wave transmittance increases the thinner they become.

Rectangular-celled extruded polycarbonate honeycombs from various plastics have been developed, but only some of them are on the market. Commercial products of small-celled honeycomb with improved optical and thermal properties were developed in the early 1980s by companies such as Diveno and Okalux Kapillargles GmbH in Germany. These products use PMMA (polymethylmethacrylate) and PC (polycarbonate) materials, which also provide ultraviolet stability behind a cover glass.

The heat transfer in large-celled honeycombs has been investigated by a number of researchers (Symon and Peck 1983, Hoogendoorn 1985 and Hollands *et al.*

1984, 1992). This structure guarantees low transmission losses in the solar spectrum and permits easy combination with selective coatings. Meanwhile, different shapes of honeycomb structures have been investigated for heat losses and solar transmission such as sinusoidal (McMurrin and Buchberg 1981), hexagonal (Marshall *et al.* 1976) and cylindrical shape honeycomb (Buchberg *et al.* 1976). For applications dealing with daylighting and light-guiding, the material is used also in modified forms. The honeycomb cells need not be perpendicular to the glazing surface, but may also be inclined, usually 45°. Also, honeycombs parallel to the glass sheets are used.

The activities in testing and monitoring TI applications have increased significantly since the late 1980s. Braun *et al.* (1992) reported 20 projects of transparently insulated walls, most of them honeycomb type, granted by the German Ministry for Research and Technology. In 1986 the European Commission Directorate General XII for Science, Research and Development launched the PASSYS project to develop reliable and affordable procedures for the testing of thermal and solar characteristics of passive solar components in a building system. The TIM component, honeycomb structured, was tested in five countries – Belgium, Germany, Italy, the United Kingdom and the Netherlands (Vandaele and Wouters 1994). The largest demonstration project was the students' residence at Strathclyde University in England (Twidell *et al.*1994), where 1040 m^2 of TIM were applied.

3.3.1 Design of honeycomb cell

The main purpose of the honeycomb is to reduce the free convection currents that would otherwise occur in the air layer. Once the value of honeycomb thickness has been decided, the cell size must be chosen so as to just suppress the free convective currents at the value of design ΔT (the temperature difference between the hot and cold faces of the honeycomb). The use of smaller cell sizes will result in more expensive honeycomb being used than is actually necessary and will also cause unnecessary loss in solar transmission of the honeycomb. However, as Cane *et al.* (1977) observed, the use of cell sizes larger than needed may induce a convective heat exchange greater than that which would take place without the honeycomb. If the aspect ratio is less than 2, very little convection suppression is possible and it is difficult to avoid the region of free convection augmentation by honeycomb. Consequently, a honeycomb with aspect ratio less than 2 should be avoided. The free convection heat transfer across a honeycomb can be characterized by the Nusselt number, Nu, which represents the ratio of the combined conductive and free convective heat transfer through the air to the corresponding heat transfer when the air is stagnant (Holman 1986). Hence, a Nusselt number of unity indicates purely conductive heat transfer and complete suppression of convective motion.

On the other hand, as observed by Charters and Peterson (1972) through flow visualization tests, there is never a complete suppression of convection motion for an inclined cell. Thus, the Nusselt number is never exactly unity even at very small Rayleigh numbers (*Ra*). However, the Nusselt number increases rather slowly with *Ra* until it reaches approximately 1.2 after which it rises rather steeply with *Ra* (Cane *et al.* 1977). Consequently, when $Nu = 1.2$ is reached, the effective suppression will be assumed to have taken place where purely conductive heat transfer of still air is augmented by 20%.

Cane *et al.* (1977) reported an extensive experimental study on free convection across inclined square and hexagonal celled plastic honeycomb. They recommend the following for Nusselt number:

$$Nu = 1 + 0.89\cos(\beta - 60)\left(\frac{Ra}{2420\,A_{hc}^{4}}\right)^{2.88 - 1.64\sin\beta} \qquad (3.20)$$

where β is the tilt angle (angle of inclination for honeycomb panel in degrees) and A_{hc} is the aspect ratio of the cell.

This relation is valid when $30° \leq \beta \leq 90°$ and $Ra/A_{hc}^{\ 4} \leq 6000$.

Although based on data for $A_{hc} \geq 4$, this relation is approximately valid for $A_{hc} = 3$ as well (Cane *et al.* 1977).

Substituting $Nu = 1.2$ in equation 3.20:

$$A_{hc} = f(\beta)(Ra/2420)^{1/4} \qquad (3.21)$$

where $$f(\beta) = \left[4.45\cos(\beta - 60)\right]^{\frac{1}{11.52 - 6.56\sin\beta}}$$

for $30° \leq \beta \leq 90°$.

For air at atmospheric pressure and moderate temperatures equation 3.21 can be expressed dimensionally as:

$$A_{hc} = C(\beta)(1 + 2x)^{1/2} \times \Delta T^{1/4} L^{3/4} \qquad (3.22)$$

where $C(\beta) = 1.03f(\beta)$ and $x = 100/T_m$. T_m is the mean air temperature expressed in kelvin, ΔT is the temperature difference across the honeycomb and L is the depth of the honeycomb expressed in cm.

It can be seen that the cell dimensions depend on mean temperature and temperature difference. However, as shown by Hollands (1978), with a good factor of safety, a cell size of the order of 12 mm in hydraulic diameter is appropriate for a cell length of 100 mm.

In choosing the cell size, the honeycomb designer must first make a decision choice concerning the chief function of the honeycomb, and this differentiation on the basis of function leads to a calculation of honeycomb into what may be called 'large-celled' and 'small-celled' honeycombs. Another classification may be governed by the cell geometry and shape. However, when met with the concept of honeycomb, a host of questions arise concerning the best geometry, dimensions and shape of the cell. In fact, the amount of the material that must go into making the transparent insulation is the key factor in answering these questions. Two factors demonstrate the importance of the material content:

- the amount of absorption and scattering of solar radiation inside the honeycomb wall is proportional to the material content, as is the heat transfer by conduction through the cell
- the material cost.

Reducing the amount of material may be achieved by reducing the thickness of the walls and increasing the cell dimensions. However, a decision should be made whether the cell is designed to suppress radiative or convection heat transfer. Long-wave radiation suppression needs a small cell and a minimum thickness of wall to establish the required degree of opaqueness. On the other hand, for convection suppression, a minimum cell size is required below which no added advantage is gained. As designing to suppress radiant heat transfer is more exacting, another method may be applied. It is the use of a selective surface on the absorber to help the radiant suppression. However, the cost of the coating has to be considered and an air gap should be added between the absorber and the TI element to decouple the conductive and radiative heat transfer mode.

3.3.2 Solar transmittance of transparent honeycomb insulation

The successful use of honeycomb transparent insulation for solar energy application depends on not only cost, convection suppression and thermal radiation reduction considerations but also on its transmittance to incoming solar radiation over different angles of incidence (Symons 1982). The solar collection effectiveness of a solar device can be judged by an accurate determination of its solar transmittance. Theoretical studies for such determinations have been made using methods such as Monte Carlo ray-tracing techniques (Morris 1976, Feland 1978). The complexities of this technique have hindered its use. Hollands *et al.* (1978) derived an approximate equation for honeycomb transmittance. However, as shown by Jesch *et al.* (1987), the values calculated by this method deviate from experimental data after 60° angle of incidence. Symons (1982) suggested empirical solar transmittance correlation equations based on a simple convection suppression solar transmittance model. The results match the measured data to within 2%. Later, Chattha and Jesch (1989) used and validated this method.

3.3.3 Beam transmittance of honeycomb transparent insulation

Symons (1982) reported a simple set of formulations for calculating the solar transmittance of transparent honeycombs. Chattha and Jesch (1989) have used these formulations for investigating the variation of solar transmittance with incidence angle and included the effect of scattering and absorption of radiation by vertical walls.

Solar radiation falling on the top surface of a honeycomb propagates through the cell as well as through the cell walls. The portion propagating through the cells undergoes reflection (specular as well as diffuse), refraction and absorption by the vertical walls.

The radiation of the cell walls is propagated downwards through the walls because of total internal reflection and, in the process, is absorbed and scattered inside the material of the walls. Following Symons (1982) and Chattha and Jesch (1989), the beam radiation transmittance $T(\theta)$ can be approximated as:

$$T(\theta) = (\tau_s + \rho_s)^{nw} \tag{3.23}$$

where nw is the number of walls intercepting incoming rays at angle of incidence θ and azimuth ϕ and is equal to $A_{hc}\tan(\theta)$ for square celled honeycomb.

For smooth wall materials, the sum of specularly reflected and transmitted radiation is essentially independent of angle of incidence. For example, Symons (1982) evaluated the value of $\tau_s + \rho_s$ for FEP Teflon honeycomb as approximately equal to 0.983. He measured the variation of this value with the angle of incidence and concluded a reduction of less than 1% when the angle of incidence increases from 0 to 75°. On the other hand, equation 3.23 is only applicable for very thin honeycomb wall materials. It can shown from equation 3.24 that for θ equal 0, $T(\theta)$ is equal to 1.

Therefore, equation 3.23 should be modified to accommodate thicker and wavy honeycomb wall. It may be expressed as:

$$T(\theta) = F_e F_r^{A_{hc}\tan(\theta)} \tag{3.24}$$

where F_e is the fraction of effective cross-sectional area of honeycomb not occupied by wall material. That is (Kaushika and Padma Priya 1991)

$$F_e = 1 - \frac{\delta_{hc}(\delta_{hc} + 2D_{hc})}{D_{hc}^2} \tag{3.25}$$

F_r is the fraction of radiation that is reflected and transmitted specularly at each wall intercept, and it is measured experimentally taking into account the absorption and scattering of solar radiation. Hollands *et al.* (1978a), Kaushika and Padma

Priya (1991) and Chattha and Jesch (1989) reported measured values for different types of honeycomb. These values are summarized in Table 3.3.

Table 3.3 F_r and δ_{hc} for different honeycomb materials

	Acrylic	*Polycarbonate*	*Glass*	*Lexan*	*FEP Teflon*
F_r	0.963	0.988	0.955	0.999	0.999
δ_{hc}	1 mm	0.03 mm	1 mm	0.076 mm	0.013 mm

3.3.4 Solar diffuse radiation transmittance of transparent honeycomb

Solar diffuse radiation transmittance for the plane glass cover of a flat plate collector has been investigated by Brandemuehl and Beckman (1980). Their approach involves the consideration of solar diffuse radiation made up of two components (sky and ground reflected) to be hemispherical isotropic in nature. For the derivation of transmittance of diffuse radiation the results of beam radiation transmittance are integrated over the appropriate range of the solid angle of incident diffuse radiation. The effective beam incidence angle θ_e is defined for a given range and transparent system such that

$$T_d = T(\theta_e)$$

The same approach may be applied to cellular array (honeycomb structured) materials. The transmittance for solar diffuse radiation is expressed as:

$$T_d = \frac{\int_{\omega 1}^{\omega 2} T(\theta)\cos(\theta)\sin(\theta)\mathrm{d}\phi\mathrm{d}\theta}{\int_{\omega 1}^{\omega 2} \cos(\theta)\sin(\theta)\mathrm{d}\phi\mathrm{d}\theta} \tag{3.26}$$

where $\omega 1$ and $\omega 2$ represent the angular range of incident diffuse radiation, and θ and ϕ are the angle of incidence and azimuth respectively. The angular range has been determined earlier by Brandemuehl and Beckman (1980) and the results are adopted in the present work.

For honeycomb transparent material tilted at an angle β, the diffuse radiation from the ground will have an angle of incidence ranging from $\theta = \pi/2 - \beta$ to $\theta = \pi/2$. For a given value of θ the azimuth angle will range from $\phi = \text{asin}[\cot(\beta)/\tan(\theta)]$ to $\phi = \pi - \text{asin}[\cot(\beta)/\tan(\theta)]$.

Using these integration limits and due to axial symmetry, the transmittance for diffuse ground radiation is:

$$T_{\text{dground}} = \frac{\int_{\frac{\pi}{2}-\beta}^{\frac{\pi}{2}} \int_{\operatorname{asin}\left[\frac{\cot(\beta)}{\tan(\theta)}\right]}^{\frac{\pi}{2}} T(\theta)\cos\theta\sin\theta\,d\phi\,d\theta}{\int_{\frac{\pi}{2}-\beta}^{\frac{\pi}{2}} \int_{\operatorname{asin}\left[\frac{\cot(\beta)}{\tan(\theta)}\right]}^{\frac{\pi}{2}} \cos\theta\sin\theta\,d\phi\,d\theta} \tag{3.27}$$

Diffuse radiation from the sky will have angle of incidence ranging from $\phi = 0$ to $\phi = \pi/2$. For the incidence angle $\phi < \pi/2 - \beta$, the angle ϕ will vary from 0 to 2π and for the incidence angle $\phi > \pi/2 - \beta$ the angle ϕ will range from $\phi = \pi - \operatorname{asin}(\cot(\beta)/\tan(\theta))$ to $\phi = 2\pi + \operatorname{asin}(\cot(\beta)/\tan(\theta))$. By applying these integration limits and due to axial symmetry the sky diffuse transmittance for a transparent honeycomb is expressed as:

$$T_{\text{dsky}} = \frac{\int_0^{\frac{\pi}{2}} \int_{\frac{\pi}{2}}^{\frac{\pi}{2}} T_b(\theta)\cos(\theta)\sin\theta\,d\phi\,d\theta + \int_{\frac{\pi}{2}-\beta}^{\frac{\pi}{2}} \int_{\frac{\pi}{2}}^{\operatorname{asin}\left[\frac{\cot(\beta)}{\tan(\theta)}\right]} T_b(\theta)\cos\theta\sin\theta\,d\phi\,d\theta}{\int_0^{\frac{\pi}{2}} \int_{\frac{\pi}{2}}^{\frac{\pi}{2}} \cos(\theta)\sin\theta\,d\phi\,d\theta + \int_{\frac{\pi}{2}-\beta}^{\frac{\pi}{2}} \int_{\frac{\pi}{2}}^{\operatorname{asin}\left[\frac{\cot(\beta)}{\tan(\theta)}\right]} \cos\theta\sin\theta\,d\phi\,d\theta} \tag{3.28}$$

Arulanantham and Kaushika (1994) used numerical integration techniques to solve the above equations. The results were reduced in terms of equivalent beam angle of incidence and its variation as a function of tilt angle and aspect ratio of the transparent device. The variations are represented as polynomial expressions. The polynomial representation is based on the theoretical calculation covering the aspect ratio range from 2 to 40.

Example

Determine the thermal and solar properties of a square-celled Lexan honeycomb with the properties given below. First let us consider the appropriate properties of the transparent material used (Lexan square-cell honeycomb). The calculations are performed in Mathcad (MathSoft 2000).

$L_{hc} = 47.6$ mm	Honeycomb depth (height)
$\delta_{hc} = 0.076$ mm	Honeycomb wall thickness
$D_{hc} = 9.5$ mm	Honeycomb wall spacing
$E = [\delta_{hc}(\delta_{hc} + 2D_{hc})]/D_{hc}^2$	Fraction of cellular cross-section area occupied by wall material (for a square cell)

$\rho_{se} = 0.97\ (\Phi = 0)$ Equivalent specular reflectivity of honeycomb wall

$\rho_{de} = 0.02\ (\Phi = 0)$ Equivalent diffuse reflectivity of honeycomb wall

$\alpha_e = 1 - \rho_{se} - \rho_{de}$ Equivalent absorptivity of honeycomb wall ($\alpha_e = 0.01$)

$\alpha_s = \alpha_e$ Absorptivity of single-sheet honeycomb

$F_{sq} = ½$ Hottel's shape factor for radiative exchange

$N_{hc_i} = \frac{L_{hc}}{D_{hc}} \tan(\theta_{i,0})$ The number of walls intercepted by the solar ray in its propagation through the honeycomb array (for square cell)

$nt_i = \text{floor}(N_{hc_i})$ Truncated value of N_{hc}

$T_{d_i} = \rho_{se}{}^{nt_i}(nt_i + 1 - N_{hc_i}) + \rho_{se}{}^{nt_i+1}(N_{hc_i} - nt_1)$ Fraction of irradiation on the top of the cell which arrives at the bottom of the cell via specular reflections only

$T_{c_i} = T_d + \left[\frac{\rho_{de}}{\rho_{de} + \alpha_s}\right] F_{sq}\left(1 - T_{d_i}\right)$ Honeycomb cell transmittance

$\mu = 1.5$ Index of refraction of honeycomb wall

$R_{sv} = (1 - \mu)^2/(1 + \mu)^2$ Single surface reflectivity at vertical incidence

$$a_{\Phi 0} = \frac{\rho_{se} - R_{sv}}{1 - 2R_{sv} + R_{sv}\rho_{se}}$$

$$a(\Phi) = \exp\{-\mu(K_a + K_s)\delta/\mu^2 - [\sin^2(\Phi)]^{1/2}\}$$

Let $K_a + K_s = K_b$, where K_s and K_a are the scattering and absorption coefficient of honeycomb and wall, respectively. This will lead to the following formula:

$$K_b = \frac{\log(a_{\Phi 0})\left(\mu^2 - \sin(0)^2\right)^{\frac{1}{2}}}{-\mu \cdot \delta_{hc}} = \log(a)/\delta_{hc}$$

$$b_{\theta_i} = \frac{K_b \mu L_{hc}}{\left(\mu^2 - \sin(\theta_{i,0,0})^2\right)^{\frac{1}{2}}}$$ where i refers to time interval

The transmittance through the walls of the honeycomb is then calculated, followed by the calculation of the effective transmittance, taking into account repeated reflections and transmission in the honeycomb.

The calculations are shown below and the transmittance (using Mathcad) is shown in Figure 3.12.

$\phi w_I = (\pi/2) - \phi_{i,0}$ Angle of incidence of solar radiation on honeycomb wall

$rs_i = \text{asin}[\sin(\phi_{i,0})/\mu]$ $\sin(rs) = \sin\theta/\mu$

$$R_{\theta_i} = \frac{1}{2}\left[\frac{\tan(\theta_{i,0} - rs_i)^2}{\tan(\theta_{i,0} + rs_i)^2} + \frac{\sin(\theta_{i,0} - rs_i)^2}{\sin(\theta_{i,0} + rs_i)^2}\right]$$

Single surface reflectivity as a function of solar angles incidence on honeycomb top surface

$$\theta_{esradiant} = 1.007$$

let $K_u = k_s / (k_s + k_a)$ $\quad \text{rsd} = \text{asin}\left[\frac{\sin(\theta_{esradiant})}{\mu}\right]$

$$K_u = \frac{\rho_{de}}{\rho_{de} + \alpha_e} \qquad R_{\theta sd} = \frac{1}{2}\left[\frac{\tan(\theta_{esradiant} - \text{rsd})^2}{\tan(\theta_{esradiant} + \text{rsd})^2} + \frac{\sin(\theta_{esradiant} - \text{rsd})^2}{\sin(\theta_{esradiant} + \text{rsd})^2}\right]$$

$$T_{e_i} = (1 - R_{\theta_i})^2 \exp(-b_{\theta_i}) + F_{sq} K_u [1 - \exp(-b_{\theta_i})](1 - R_{\theta_i})$$ Transmittance through cell walls

$$T_{beam_i} = \frac{T_{c_i} + T_{e_i} E}{1 + E}$$

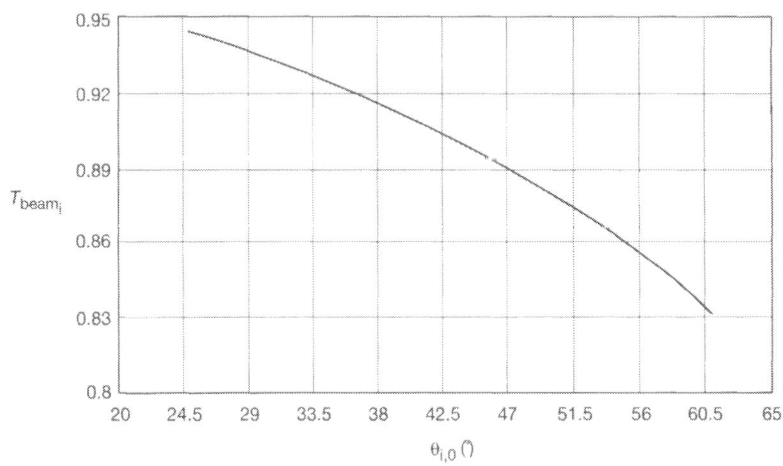

Figure 3.12 Calculation of beam transmittance of TIM (for south-facing surface in January)

3.3.5 Diffuse transmittance

Given that the aspect ratio of the honeycomb is $A_{hc} = L_{hc}/D_{hc}$ Arulanantham and Kaushika (1994) propose the following correlation and calculation technique to determine the honeycomb transmittance for ground-reflected and sky diffuse radiation.

$$\theta_{e1} = 90 - 0.6084518\beta + 3.709793\times10^{-4}\beta A_{hc} + 3.0577183\times10^{-5}\beta A_{hc}{}^2 + \left[3.30328929\times10^{-3}(\beta)^2\right]$$

$$\theta_{e2} = -5.0287694\times10^{-5}(\beta)^2 A_{hc} + 4.1570456\times10^{-7}(\beta)^2 A_{hc}{}^2$$

$$\theta_{e3} = 59.69678 - (0.3802643 A_{hc}) + \left(4.2715073\times10^{-3} A_{hc}{}^2\right) - 0.1429023\beta$$

$$\theta_{e4} = -8.0555677\times10^{-3} A_{hc}\beta + 2.1229684\times10^{-4} A_{hc}{}^2\beta + 1.527757\times10^{-3}(\beta)^2$$

$$\theta_{e5} = 7.0696505\times10^{-3} A_{hc}\beta - 2.1496235\times10^{-6} A_{hc}{}^2(\beta)^2$$

The equivalent beam angle of diffuse ground radiation is:

$$\theta_{eg} = \theta_{e1} + \theta_{e2} \qquad \theta_{eg} = 89.056 \qquad \theta_{egradiant} = \frac{\theta_{eg}\pi}{180} \qquad \theta_{egradiant} = 1.554\,\text{rad}$$

The equivalent beam angle of diffuse sky radiation is:

$$\theta_{es} = \theta_{e3} + \theta_{e4} + \theta_{e5} \qquad \theta_{es} = 57.678 \qquad \theta_{esradiant} = \frac{\theta_{es}\pi}{180} \qquad k = 1..2$$

$$b_{\theta e_k} = \frac{K_b \mu L_{hc}}{\left[\mu^2 - \sin(\theta_{k,0})^2\right]^{\frac{1}{2}}} \qquad \theta_{ek} = \theta_{egradiant} / \theta_{esradiant}$$

$$T_{e_k} = (1 - R_{sv})^2 \exp(-b_{\theta_k}) + F_{sq} K_u \left[1 - \exp(-b_{\theta e_k})\right](1 - R_{sv})$$

$$N_{hc_k} = \frac{L_{hc}}{D_{hc}} \tan(\theta_{k,0}) \qquad nc_k = \text{ceil}(N_{hc_k})$$

$$T_{d_k} = \rho_{se}^{nc_k}(nc_k + 1 - N_{hc_k}) + \rho_{se}{}^{nc_k+1}(N_{hc_k} - nc_k)$$

$$T_{c_k} = T_{d_k} + \left[\frac{\rho_{de}}{(\rho_{de} + \alpha_s)}\right] F_{sq}(1 - T_{d_k}) \qquad T_{equ_k} = \frac{T_{c_k} + T_{e_k} E}{1 + E}$$

$T_{eqground} = T_{equ_1}$ $\quad T_{equ_1} = 0.836$ $\quad$ Transmittance of ground diffuse radiation

$T_{eqsky} = T_{equ_2}$ $\quad T_{equ_2} = 0.841$ $\quad$ Transmittance of sky diffuse radiation

The beam solar absorptance of the TIM is equal to 1 minus the reflectivity and the transmittance (Figure 3.13).

Next we compute the diffuse solar absorptance of the TIM.

1 For sky radiation For $\theta = \theta_{esradiant}$

$$\text{rsds} = \text{asin}\left(\frac{\sin(\theta_{esradiant})}{\mu}\right) \qquad R_{\theta sd} = \frac{1}{2}\left[\frac{\tan(\theta_{esradiant} - \text{rsds})^2}{\tan(\theta_{esradiant} + \text{rsds})^2} + \frac{\sin(\theta_{esradiant} - \text{rsds})^2}{\sin(\theta_{esradiant} + \text{rsds})^2}\right]$$

$R_{\theta sd} = 0.079$ $\quad \alpha_{skydiffuse} = 1 - (R_{\theta sd} + T_{eqsky})$ Sky diffuse absorptance

2 For ground radiation

$$\text{rsdg} = \text{asin}\left(\frac{\sin(\theta_{egradiant})}{\mu}\right) \qquad R_{\theta gd} = \frac{1}{2}\left[\frac{\tan(\theta_{egradiant} - \text{rsdg})^2}{\tan(\theta_{egradiant} + \text{rsdg})^2} + \frac{\sin(\theta_{egradiant} - \text{rsdg})^2}{\sin(\theta_{egradiant} + \text{rsdg})^2}\right]$$

$R_{\theta gd} = 0.909$ $\quad \alpha_{grounddiffuse} = 1 - (R_{\theta gd} + T_{eqground})$ Sky diffuse absorptance

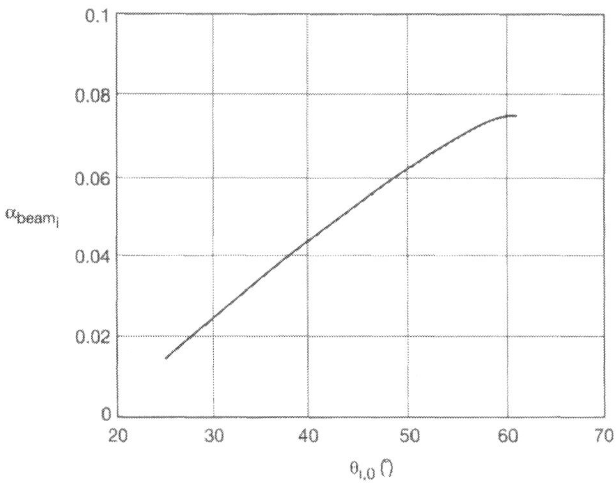

Figure 3.13 Calculation of absorptance of TIM (in Mathcad)

3.4 Shading

Shading provision should be considered as an integral part of fenestration system design. Blocking the sun before it reaches the building interior is the most effective method of reducing cooling loads.

The design of a fixed shading system depends on the opening area and the orientation of the aperture with respect to the position of the sun. In general, horizontal shading is used in south facades, while vertical is more efficient on the east or west facade.

The majority of fixed shading systems are installed in the exterior of a building in order to dissipate the heat that they absorb from the sun to the outside air. The same device used as internal shading will result in an average 30% reduced efficiency. This is because the shading devices dissipate heat by convection and long-wave radiation. Therefore, internal devices dissipate this heat inside the building space, since convection is due to the indoor air and glazing is opaque to long-wave radiation, which is also blocked indoors.

In designing shading systems for windows there are several conflicting design requirements that must be considered before an optimal solution is reached:

- blocking of direct solar gains during the cooling season
- allowing the maximum possible amount of solar radiation into the building interior during the heating season
- control of the intense daylight during clear days by diffusing it into the space or reflecting part of it towards the ceiling for example
- to allow the possibility of simultaneous shading and natural ventilation.

Shading of fenestration components may be achieved using fixed or movable devices (Figure 3.14). Fixed shading devices include:

- continuous overhangs, or horizontal louvres above the window
- side fins, or vertical louvres on the sides
- awnings (fixed in summer, retractable in winter).

Louvres, when strategically placed, allow winter sun to pass between their gaps, while blocking it in summer.

The window is also usually recessed into the wall (with the exception of curtain walls) resulting in the same shading effect as side fins and overhang of the same dimension.

Movable shading devices for fenestration systems may be external or internal. They include:

- roller shades (retractable)
- venetian blinds
- curtains (usually internal).

Adjustable devices may be lifted or rolled or drawn back from the window either automatically or manually. Venetian blinds are also sometimes built into double-glazing between the two glazings, and are operated by an indoor mechanism.

Architectural projections, known as **overhangs** when they are horizontal or **side fins** when vertical, are commonly employed in the building envelope to control the amount of solar radiation reaching a surface. The shading device is typically designed to exclude solar radiation in summer when it contributes to a reduction in cooling loads, and to admit it in winter when it contributes to a pleasant indoor environment and a reduction of heating requirements.

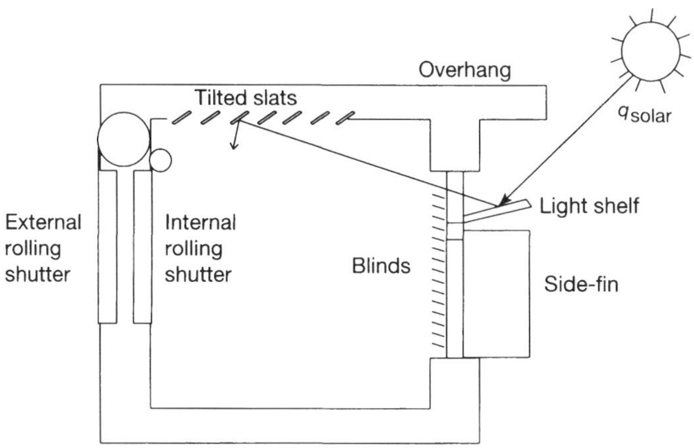

Figure 3.14 Shading devices

Shading of windows by exterior or adjacent components may be computed using basic solar geometry and projection equations, computer graphics techniques or graphical techniques.

A geometric technique for finding the shaded area of a recessed window with an overhang is based on Figure 3.15. A window of dimensions $c \times a$ recessed at a distance b from the wall outside surface is shown. The objective is to design an overhang with length equal to $c + 2g$ and width f, at a distance e above the window. The ray projection diagram shows the minimum dimensions for complete shading of the window for the given solar altitude α, surface solar azimuth angle γ and surface azimuth angle ψ. Point E at the outer edge of the overhang is projected to point O on the window plane.

The profile angle d is given by

$$\tan d = \tan \alpha / \cos \gamma \tag{3.29}$$

It can be shown that the geometric dimensions of the overhang that satisfy our constraints are given by:

$$f = [(a + e)/\tan d] - b$$
$$g = f \tan \gamma \tag{3.30}$$

Example

Consider a south-facing window located in Athens. The design objective is to shade the window from 9 a.m. to 3 p.m. solar time on 15 May. Because it is south-facing, 9 a.m. and 3 p.m. are symmetrical times and only 9 a.m. needs to be considered.

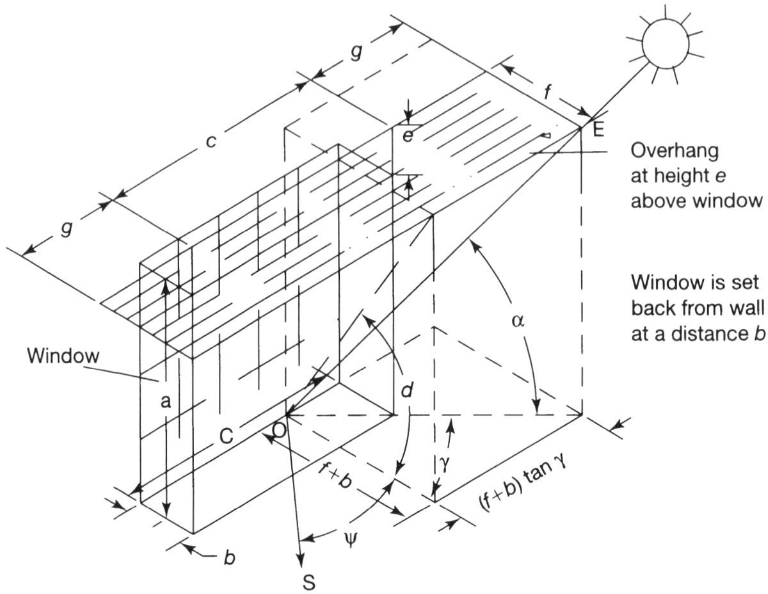

Figure 3.15 Recessed window with an overhang

Window geometry:

$a = 1.5$ m $c = 2.0$ m $b = 0.2$ m e $= 0.2$ m

$L = 38°$ $\varphi = 0$ $\beta = 90°$

$\delta = 18.8°$ (declination)

$\alpha = 46.6°$ (solar altitude)

$\varphi = \gamma = -76.7$ (solar azimuth and surface solar azimuth)

Profile angle $d = 77.7°$

Overhang dimensions:

width $f = 0.17$ m, $g = 0.72$ m, i.e. length $c + 2g = 3.44$ m

If these dimensions are chosen, the window will be completely shaded from 15 May to 28 July (symmetrical dates about 21 June).

It is interesting to check how much of the window is shaded at solar noon on 21 December. In this case the solar altitude is 24.5° and it is also equal to the profile angle d. The projection of the overhang edge on the window plane lies at a distance a' from the window upper edge:

$$a' = (f + b) \tan d - e$$

$$a' = 0.002 \text{ m}$$

Therefore only 0.2% of the window is shaded.

Overhangs are appropriate for near-south-facing surfaces while side fins shade better east- and west-facing surfaces. For a side fin (Figure 3.16) of width D the shaded part of the window h can be estimated as follows:

$$h = D \times \tan(\text{solar azimuth} - \text{window azimuth})$$

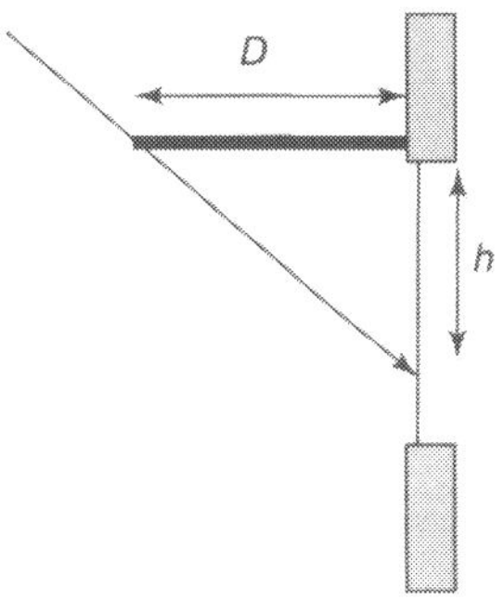

Figure 3.16 Shading of a window with a tall side fin of width D

Graphical techniques may also be employed for determination of the position of the sun and shading estimation.

A simplified method to determine the mean monthly shading coefficient has been developed under the PASSYS (1994) programme and is implemented in the PASSPORT computer program developed for the EC in 1994. The method considers three types of shading device:

- recessed window
- side fins
- overhangs.

All the geometric dimensions are input to PASSPORT, which computes the mean monthly incident solar radiation based on the following expression:

$$H_{b,g} = (1 - H_d/H)R_b H \, \mathrm{SB}_1 \, \mathrm{SB}_2 + H_d(1 + \cos b)/2\mathrm{SD} + rH(1 - \cos b)/2\mathrm{SR} \quad (3.31)$$

where:
SB_1 = direct radiation shading factor for remote obstacles

SB_2 = direct radiation shape factor for facade obstacles
SD = diffuse radiation shading factor
SR = reflected radiation shading factor

$$SB_1 = (h_{ss} - h_{sr})/(2h_s)$$

h_{ss}, h_{sr} = sunset and sunrise times corresponding to the average angle of the remote obstacle

$$2h_s = 2(\text{sunset hour angle}) = \text{daylength}$$

For a recessed window SB_2 is determined by

$$SB_2 = \exp[(A_1 + B_1(L - \delta))R/H + (A_2 + B_2(L - \delta))R/W] \tag{3.32}$$

where W is the window width and R the depth of the opening. The constants A_1, B_1, A_2 and B_2 were calculated from detailed simulations and are given in Table 3.4. Similarly, SB_2 for side fins is given by

$$SB_2 = 1 + [A_1 + B_1(L - \delta)]l/W + [A_2 + B_2(L - \delta)](l/W)(D/W) \tag{3.33}$$

where:
l = length of side fin
D = distance between window and side fin

Table 3.4 Constants for monthly shading calculation [Santamouris 1994]

Shading component	*Direction*	A_1	B_1	A_2	B_2
Recessed window	S	–5.695	0.081	–1.342	0.009
	SE, SW	–2.418	0.032	–1.479	0.017
	E, W	–0.868	0.009	0.232	–0.022
	NE–NW	0.336	–0.013	–0.320	–0.074
	N	–1.193	0.036	–1.825	–0.163
Side-fins	S	–1.175	0.012	0.860	–0.008
	SE, SW	–0.799	0.009	0.684	–0.006
	E, W	0.118	–0.014	0.005	0.010
	NE, NW	0.155	–0.041	0.680	0.009
	N	0.275	–0.133	0.641	0.039
Overhang	S	–3.023	0.045	1.285	–0.006
	SE, SW	–1.255	0.015	0.905	–0.008
	E–W	–0.684	0.005	0.610	–0.004
	NE, NW	–0.654	0.006	0.616	–0.006

For overhangs the parameter SB_2 is calculated with the following expression:

$$SB_2 = 1 + [A_1 + B_1(L - \delta)]l/W + [A_2 + B_2(L - \delta)](l/W)(D/W)$$

where
l = length of overhang
D = distance between window and overhang

Graphical methods may also be employed in shading analysis using sunpath diagrams and shadow charts. A simple graphical technique to visualize solar access and shading is to obtain a shadow pattern plot by finding the shadow of one shading object on the component under consideration.

3.4.1 Light shelves

The light shelf is a horizontal baffle placed at a certain height of the window opening, intended to provide shade below it and at the same time to reflect the light to the ceiling. In most cases the light shelf is fitted some distance up the window, dividing the window opening in two parts (Figure 3.17).

The benefit resulting from the use of light shelves is to increase daylight penetration deeper into the building core. Furthermore, light shelves can reduce cooling loads due to a reduction of solar gains. There are two types of light shelves, internal and external. Internal light shelves provide less daylight penetration than the external ones, under all types of sky, except when direct sunlight impinges on them. In this case they shade a portion of the space close to the aperture. However, since this type of shelf does not shade the glazing surface, cooling load reduction is negligible.

External light shelves improve daylight penetration under all sky types. The performance of light shelves varies with the reflectance of the upper and lower surface. A highly reflective material can cause glare.

Recent studies on the performance of light shelves lead to the following conclusions (Littlefair 1995):

- Light shelves do not result in a large increase of core daylight illuminances. The highest relative increases (12–20%) occur when there are large external obstructions. In most cases, small light shelves made little difference to illuminances at the back of the room. Larger light shelves tended to reduce this illumination.
- Light shelves do improve uniformity of illuminance by reducing light levels near the window.
- Light shelves provide good shading from sunlight especially in summer.
- Light shelves perform better in high ceiling spaces (>3 m).
- The reflectance of the upper light shelf surface must be as high as possible.

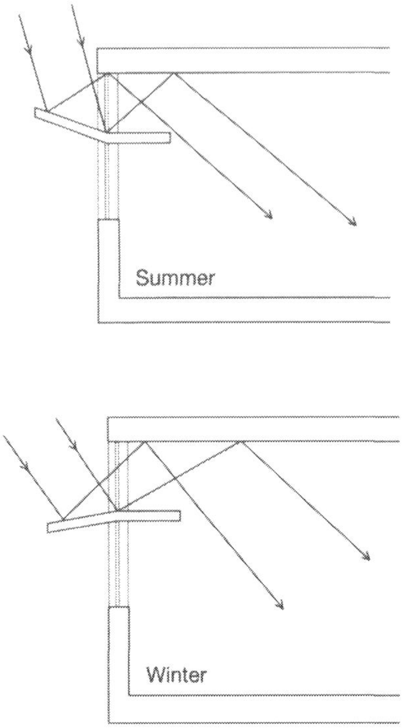

Figure 3.17 Light shelves

3.4.2 Adjustable shading devices

EXTERNALLY INSTALLED BLADES

The use of externally installed blades changes the entire appearance of the building. The lower part of the blades can be closed to avoid direct solar penetration. The upper part can be adjusted horizontally to reflect daylight deeper into the room. Daylighting levels are increased by up to 30% at 4 m distance away from the external wall. This device can be used in conjunction with inclined reflected ceiling in order to increase daylight levels on the working surface.

HORIZONTAL MOVABLE EXTERIOR LIGHT SHELF

This element can be used for shading areas near the opening and to increase daylight levels away from it. The tilt of the light shelf depends on the position, orientation and sky luminance distribution. The main disadvantage is that its efficiency is rather reduced when used under overcast sky conditions. This type of shading device is used on south facades in predominately sunny conditions.

MOVABLE AWNING

Awnings can reduce heat gains in summer by up to 65% on south facades and by up to 80% on west- or east-oriented facades. Their geometry is similar to that of horizontal overhangs, but their efficiency depends on the direct and diffuse transmittance of the material used. An additional feature of the awning is that it allows air circulation and hence natural ventilation. The transmittance of the fabric deteriorates with age. This system can be applied to almost all windows.

VENETIAN BLINDS

These are mainly internal devices, consisting of semi-transparent polyester or metal blinds. These blinds can be fixed or operable. Operable blinds can be automatic and designed in such a way as to track the sun or to track the shade. For sun tracking, they continuously change their tilt angle with respect to solar altitude to control sunlight to enter the room. For shade tracking, they change their tilt in order to reflect sunlight toward the ceiling or the exterior.

Glare adjustment can be performed by a proper choice of material. A factor for the determination of the effectiveness of a particular blind system is the ratio D/h. The parameters D and h are identified in Figure 3.18. Varying the ratio D/h from 1/1 (overhang) to 0.03/1 (thin blade) reveals three distinct phases in the penetration of daylight through the blades:

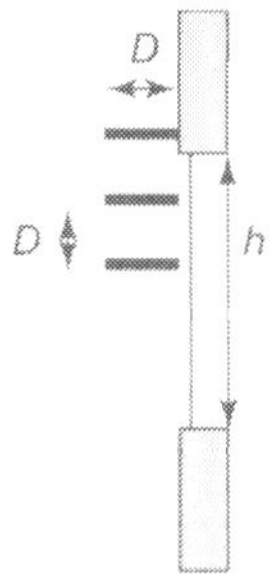

Figure 3.18 Venetian blinds

- The first phase ($1/1 > D/h > 0.5/1$) is characterized by a daylight distribution similar to that of an overhang. As the ratio of D/h approaches 0.5/1, a rapid reduction of illuminance near the window is observed, followed by a gradual improvement before the illuminance starts to drop again.
- The second phase ($0.5/1 > D/h > 0.08/1$) is characterized by a considerable variation in performance. Illuminance near the window is decreased as the ratio D/h decreases.
- During the third phase ($0.08/1 > D/h$) as the ratio D/h decreases, the performance close to the aperture increases. Performance levels drop fast at 0.01/1.

PERFORATED BLADES

These blades can be installed in the exterior or in the interior of a building as well. Depending on the perforation ratio, daylight levels can be adjusted. In general, these are not efficient for working spaces and their use is restricted to corridors, staircases and places without requirements for increased daylight.

VARIABLE-AREA LIGHT-REFLECTING ASSEMBLY

A reflective plastic film is rolled along a tilted surface so that a small steeply tilted mirror is created in summer and a larger, nearly horizontal mirror in winter. Hence sunrays are reflected to the rear of the room, both in winter and summer.

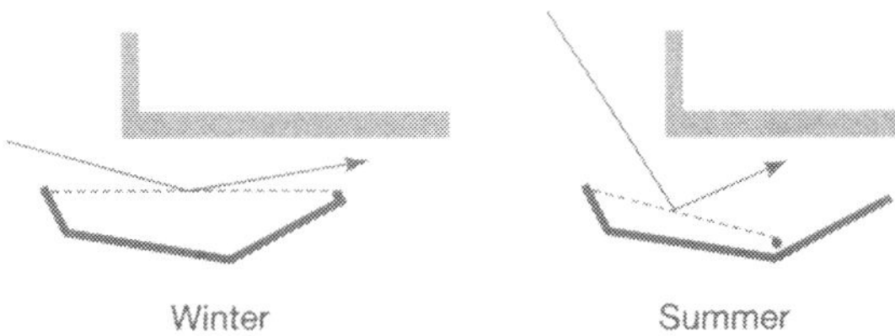

Figure 3.19 Variable-area light-reflecting assembly

Shading devices such as venetian blinds were evaluated experimentally (Aleo *et al.* 1994) for optical and thermal characteristics, and with the LAMAS software (Coronel *et al.* 1994). LAMAS is based on a detailed method for evaluation of the solar-optical performance of louvre-type shading devices. This detailed model has the following features:

- three-dimensional characterization of louvre-type devices, including different tilt angles, overlapping shades, etc.
- accurate model for direct and diffuse solar radiation
- reflection from shading elements
- semi-transparent behaviour
- coupled solar-optical performance of shading device and glazing
- coupled thermal performance of shading element and fenestration
- actual film coefficients for natural and forced convection.

Some major results of Aleo *et al.* (1994) were obtained with CFD analysis. It was found that unless the slats are horizontal the flow regime on both faces is laminar and the thickness of the boundary layer is negligible compared with the slat spacing. The correlation by Fuji and Kimura (1972) was found to be accurate for this purpose:

$$Nu = 0.56(Gr\ Pr\ \cos\theta)^{1/4} \qquad \text{for } 10^5 < Gr\ Pr < 10^{11} \tag{3.34}$$

where Pr = Prandtl number = 0.7 for air, and θ is the angle of the slat with the vertical. Nu is the Nusselt number (hL/k) and Gr the Grashof number.

Coronel *et al.* (1994) also report a wind convective heat transfer coefficient for outdoor louvre-type devices which varies from about 6 W/m^2 K for a wind speed of 0.2 m/s to about 8 W/m^2 K for a wind speed of 3 m/s. This is reduced by about 33% compared with a flat surface. To calculate the total film coefficient for outdoors, a radiative component of about 5 W/m^2 K must be added.

CURTAINS (DRAPERIES) AND FABRIC SHADES

These are another common means of internal shading. Draperies are commonly used in houses to exclude unwanted solar gain, to control glare, for thermal insulation and to provide privacy. Shading coefficients as a function of fabric reflectance, yarn reflectance and average openness are given in ASHRAE (1989) for various types of curtain weave; values are given for single and double glass types. Experiments by Pennigton (1964) gave an SC (at normal incidence) of about 0.6 for non-planar light-coloured draperies with double glazing (clear glass) and SC = 0.4 when heat-absorbing double glazing was substituted. The SC values decreased to about 0.5 and 0.33 respectively for incidence angles of 60°. Shading factors (F-factors) for double glazing with internal draperies or venetian blinds have been reported (Blanchet and Girard 1993) (Table 3.5).

Table 3.5 Shading factor F for curtains and blinds (Blanchet and Girard 1993)

Shading device	*Period of year*	*F*
Double-glazed clear	August	0.70
(4 mm sheets and 8 mm gap)		
	April	0.68
Internal drapery		
light	July	0.44
dark	August	0.65
External drapery		
light	September	0.15
dark	September	0.22
Internal venetian blind		
angle 0°	November–December	0.38
angle 45°	February	0.41
angle 90°	January	0.57

The experimental data in Table 3.5 are based on the PASSYS test procedure.

PRISMATIC SYSTEMS

Prismatic panels control transmitted light by refraction. The direction of the incoming daylight is altered by passing through a prism or a triangular wedge of glass or plastic (Figure 3.20). Further investigation of these systems has led to the development of two forms of prismatic glazing:

- sunlight-directing prisms
- sunlight-excluding prisms

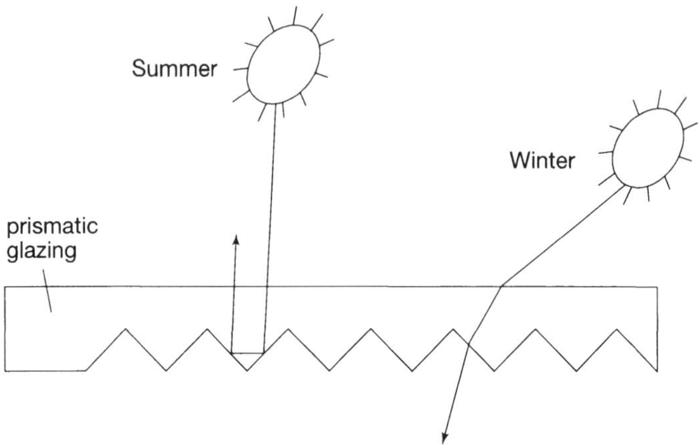

Figure 3.20 Prismatic systems

These systems when used for shading offer some advantages:

- Even though the sky cannot be seen through the systems it remains perceptible. Thus the system does not disturb the overall appearance of a window seen from the inside.
- Glare due to the sky brightness seen through a conventional window is reduced to a high degree.

3.4.3 Use of vegetation as shading device

The position and density of trees around a building contribute significantly to the reduction of solar gains. Factors that should be taken into account during the design stage of a building are the rate of growth, the size of the trees when they are mature and root development.

Basically, the luminance of one tree is a function of the illuminance on the foliage, the reflectance of the foliage, the amount of sky visible within the envelope of the crown and the sky luminance. The effective luminance of the tree crown is given by the following expression (Wilkinson 1992):

$$L_f/L_s = L_s p_s + (E_f \rho_f/\pi)(1 - p_s) \tag{3.35}$$

where L_f is the effective luminance of tree crown, L_s is the luminance of the sky behind the crown, ρ_f is the crown reflectance and p_s is the fraction of sky visible through the crown.

The crown reflectance varies with the angle of view and is less than the reflectance of individual leaves. For most purposes, the crown reflectance may be taken as 0.1 and the illuminance is taken as equal to the value on a vertical surface.

The following equation gives the luminance of a tree crown, as a function of the luminance of the overcast sky obscured:

$$\frac{L_f}{L_s} = p_s + \frac{\pi(3+7\rho_g)+8}{6\pi(1+2\sin\gamma)}\rho_f(1-p_s) \tag{3.36}$$

where ρ_f is the mean reflectance of the crown, ρ_g is the reflectance of the ground, p_s is the fraction of sky visible through the crown, and γ is the mean altitude of the sky patch hidden behind the crown. This angle can be visualized as follows: imagine the solid angle that is formed from viewpoint to the projected area of the tree crown. Draw a line that is located at the centre of the formed cone. The elevation of this line is the angle γ.

For an angle of view just above the horizon, when $\rho_g = 0.2$ and $\rho_f = 0.1$ the above equation becomes:

$$L_f/L_s = p_s + 0.1(1 - p_s) \tag{3.37}$$

This is an adequate approximation for most purposes where the reflectance and sky fraction parameters of the trees are not known in detail. Table 3.6 presents typical values of the sky cover fraction for individual mature tree crowns during summer and winter.

Table 3.6 Sky cover fraction for mature trees

	Fraction of sky visible through crown (p_s)	
Type of tree	*Summer*	*Winter*
Oak	0.17	0.5
Sycamore	0.18	0.54
Birch	0.24	0.54
Lime	0.11	0.54
Alder	0.24	0.49
Scots pine	0.27	0.3
Horse chestnut	0.19	0.52

3.5 Design of fenestration systems for daylighting

Daylighting is another important consideration and several measures may be taken to improve its quality and reduce glare. Daylighting considerations were introduced in the previous section on shading systems. Until the beginning of the 20th century, buildings were designed to exploit daylight. Then, owing to inexpensive electricity, old daylighting principles were forgotten. Now, new techniques are being developed (Kristensen 1994) and old ones are being rediscovered; see Figure 3.21. Internal reflective light shelves placed at about two-thirds height of the window reflect part of the incoming light towards the ceiling, while external overhangs situated below the top window edge may provide some shading while reflecting some of the light towards the ceiling. Skylights with shutters or electrically operated blinds may also achieve a similar effect.

Zenithal openings (Couret *et al.* 1994) are also attracting interest because they provide a high illuminance level; an anidolic toplighting consists of a compound parabolic concentrator which concentrates light in its top mirror and then diffuses it with its bottom one (Figure 3.21). It protects against direct sunlight, something difficult with atria.

Computer simulations of toplighting systems were performed with the ADELINE package, which combines heating–cooling–lighting analysis, and RADIANCE (Ward *et al.* 1988), a backward stochastic adaptive ray-tracing program. They considered diffusive glazing, a shed system (sawtooth apertures facing north) and the anidolic zenithal opening. The latter system collects skylight and guides it inside the building, always rejecting direct light. Ward *et al.* found that the zenithal anidolic lighting concept guarantees total protection against beam radiation, and without any movable parts lets diffuse skylight penetrate a large part of the building, with reduced glare.

Recent innovative solutions to diffusion of solar radiation into interior space include light-directing holographic optical elements (HOE) (Dietrich 1993). An HOE is made with a special film (laser technique); it may concentrate direct light, do a spectral division of light (like a prism), and diffuse direct light to avoid glaring effects.

Another notable technological innovation in window systems is the reversible window (Feuermann and Novoplansky 1998) in which a double-glazed window comprising a clear and a tinted glazing is rotated so that the tinted glazing is exterior in summer and interior in winter, resulting in the desirable SHGF for each season (Figure 3.21).

Another reason for avoiding direct solar radiation in office buildings is its strong effect on the effective temperature felt by persons in its path. Athienitis and Haghighat (1992) report that the effect of transmitted solar radiation on 'effective globe temperature' for double-glazed windows can be about 10°C (see Section 8.1).

Windows are unique from the occupant's perspective because they provide light and view, functions not provided by other elements. From the designer's point of view they are unique because they represent a significant element in the

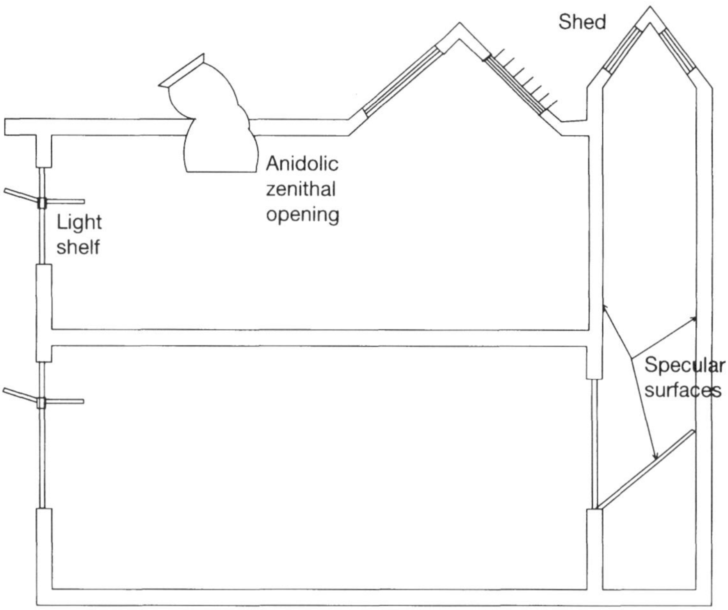

Figure 3.21 Daylighting techniques

vocabulary of architectural design. As such they are often used as formal elements of architectural expression in addition to the energy and non-energy performance requirements discussed previously.

The window is a dynamic element in the building envelope and must respond to continuously changing occupant needs as well as continuously changing environmental conditions.

A large portion of the energy consumed by buildings is due to lighting. There are many ways of saving lighting energy, but once again these solutions must be consistent with maintaining or improving productivity. Despite the high cost of energy, the value of human productivity is many times greater. In North America, the annual cost of providing lighting energy for a small office occupied by a single occupant is approximately equal to that worker's salary for a single hour. Thus, if a poor lighting design costs the employer even one hour's worth of productivity in return for large annual lighting savings, the employer has lost money overall.

A complete discussion of daylighting in buildings would consider its impacts on the following issues: electric lighting integration, energy savings, peak load impacts, HVAC systems impacts, lighting quality, view and glare. As an architectural design element, daylighting influences the built environment at many different scales: urban planning, building form, envelope design, fenestration design and interior design.

There are significant differences between using sun and sky as a source of light and using electric light sources. A primary difference is the inherent variability in daylight and its unpredictability. Standard clear and overcast skies have been defined and used for some time by the international lighting community:

- **Uniform luminance sky distribution**. This model represents a sky with constant value of luminance (in this case the resulting horizontal illuminance is πL).
- **CIE overcast sky distribution**. In this model the luminance is varied with elevation and corresponds to a situation where the sky is covered with light clouds and the sun is not visible (Figure 3.22). The luminance in a point (L) of the sky with elevation θ is given by:

  $$L = L_z(1 + 2\sin\theta)/3 \tag{3.38}$$

  where L_z is the zenith luminance. The transfer process of the radiation through clouds, in the absence of severe pollution, produces white light by colour mixing. If the atmosphere is heavily polluted, the overcast sky colour appearance is yellow.
- **Clear-sky luminance distribution**. This type of sky has a strongly non-uniform luminance distribution. The CIE has recommended standard luminance distributions of the clear sky which are simplifications of a complex process of atmospheric scattering in the presence of different pollutants. Most clear-sky luminance models are variants of the same basic form. The formula presented below relates the luminance L in one point of the sky vault with the zenith luminance L_z:

(a) (b)

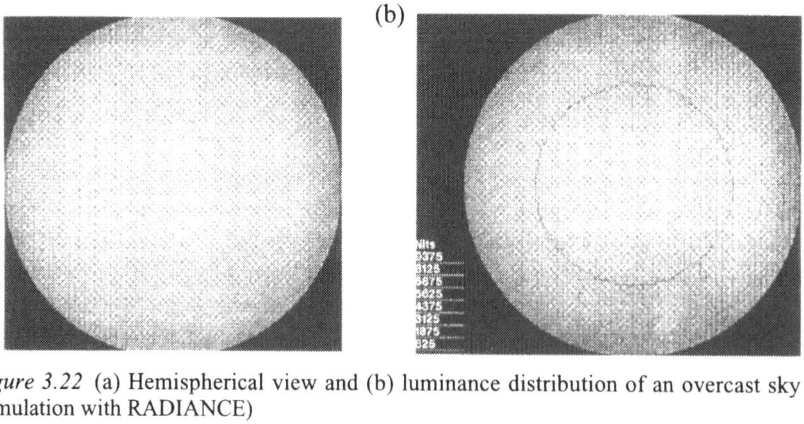

Figure 3.22 (a) Hemispherical view and (b) luminance distribution of an overcast sky (simulation with RADIANCE)

$$\frac{L}{L_z}=\frac{\left(a+b_{e}^{-3\theta}+c\cos^{2}\theta\left(1-e^{-d\sec z}\right)\right)}{\left(a+b_{e}^{-3z_s}+c\cos^{2}z_s\right)\left(1-e^{-d}\right)} \qquad (3.39)$$

where a, b, c, d are adjustable coefficients, θ is the angle (in radians) between the point with luminance L and the sun, z is the zenith angle of the point and z_s is the zenith angle of the sun.

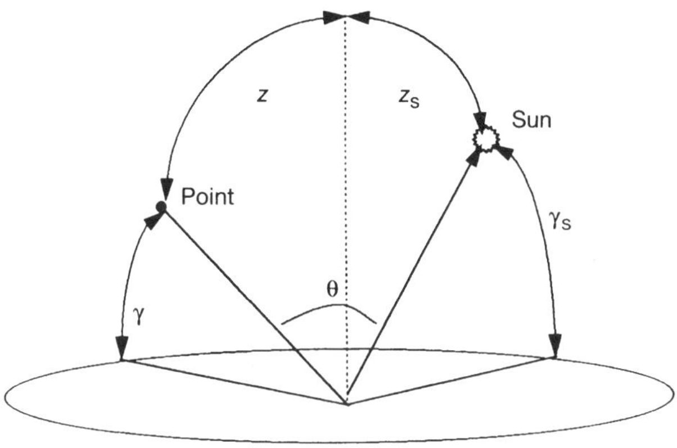

Figure 3.23 Definition of parameters used in sky luminance distribution functions.

The CIE clear sky model has a = 0.91, b = 10, c = 0.45, d = 0.32. Additionally CIE has also standardized a clear sky for polluted atmospheres, using the following coefficients: a = 0.856, b = 16, c = 0.3, d = 0.32 (Figure 3.23).

The blue colour of the clear sky is strongly dependent on the height of the site above sea level and by the amount of atmospheric pollution. Nitrogen dioxide makes the colour of the atmosphere brown and this can be seen when looking towards an urban area from the surrounding countryside. High water vapour levels, in the absence of pollution, tend to give the sky a white appearance (Figure 3.24).

However, these standards were often created with minimum conditions in mind, in order to verify a minimum design standard. For the purposes of either estimating occupant satisfaction or calculating energy effects, these sky models must be expanded to include partly cloudy and direct sun effects and to account for the variability in daylight levels and sky luminance distribution over the entire year. Research is under way in many locations worldwide to fill these gaps.

To characterize the way in which daylight penetrates a building, it is useful to examine the distribution of indoor illuminance as a function of the external luminous conditions. Owing to the simplicity of the luminous distribution function of the overcast sky, the interior illuminance in a point of a building is proportional to the external horizontal illuminance. This ratio of the daylight illuminance at an

(a) (b)

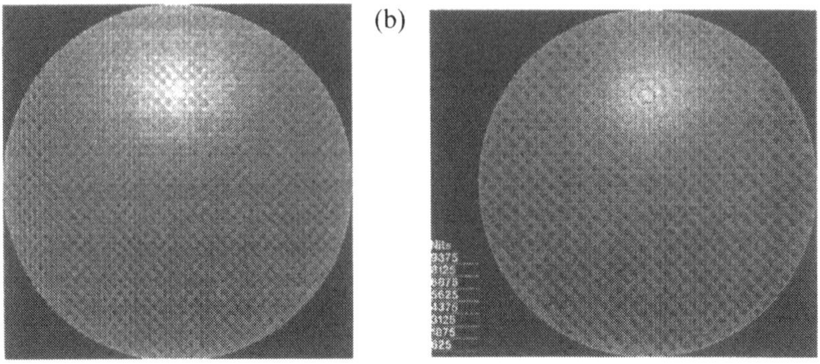

Figure 3.24 (a) Hemispherical view and (b) luminance distribution of a clear sky (simulation with RADIANCE)

interior point to the instantaneous illuminance outside the building on a horizontal surface from a complete hemisphere of sky is called the **daylight factor** and can be expressed as a percentage:

$$DF = E_{in}/E_{ext} \times 100 \tag{3.40}$$

Thus the distribution of the indoor illuminance can be carried out by the use of daylight factors (i.e. having as reference the overcast sky). Under clear sky conditions the above-mentioned concept of daylight factor cannot be used as a design criterion since it varies significantly throughout the day.

A convenient way to evaluate the total amount of daylight in a room and hence the room brightness in relation to the view outside is the average daylight factor. The working surface is considered to be a horizontal surface at 0.7 m above the floor for offices. The estimation of average DF is based on the calculation of total luminous flux entering the window, the ratio of window opening to internal surface and the overall interreflections.

The calculation of the average daylight factor with simple external obstruction can be done as follows.

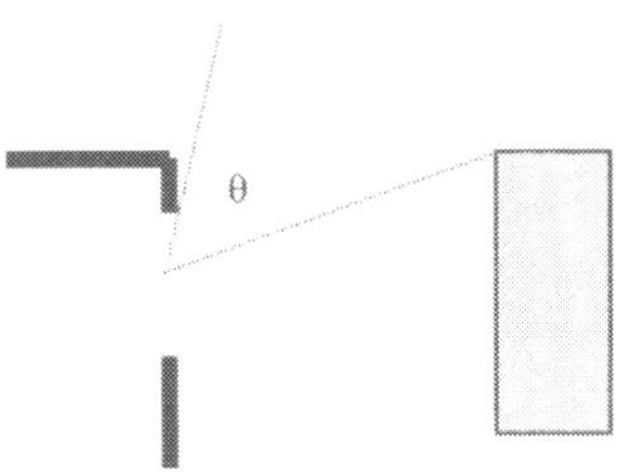

Figure 3.25 Definition of obstruction angle

$$DF = \tau\{A_w\theta/[A(1 - \rho_{mean})]\} \quad (3.41)$$

where τ is the diffuse light transmittance of the glazing, A_w is the area of the window in m^2, A is the total area of ceiling, floor and walls including windows in m^2, θ is the angle subtended by the visible sky (Figure 3.25), measured from a point in the plane of the internal wall at the centre of the window, in degrees and ρ_{mean} is the area-weighted reflectance of ceiling, floor and walls including the windows. However, during the design process it is useful to investigate how daylight arrives at a point inside a room. This can be done by splitting the available indoor illuminance into three components:

- Sky component (DF_{SC}), which is the illuminance received at a point in the interior of a building, directly from the sky.
- External reflected component (DF_{EXT}), which is the illuminance in the interior due to light reflected from external obstructions.
- Internal reflected component (DF_{INT}), which is the illuminance received at a point and is composed of light received indirectly from daylight that is interreflected around the internal surfaces of the space.

Therefore, for any point indoors:

$$DF = DF_{SC} + DF_{EXT} + DF_{INT} \quad (3.42)$$

It follows therefore that if no sky is visible from the point, $DF_{SC} = 0$. Further, if no window can be seen from the point , $DF_{SC} = DF_{EXT} = 0$.

For over 80% of the floor area in most buildings daylight factors fall in the range of 0–5%. Close to the windows, however, they can reach 10–15%. The values of the daylight factors, therefore, provide a guide to the brightness achieved in a room from daylight. A classification of the brightness of building zones based on the values of daylight factors is given in Table 3.7.

Table 3.7 Average daylight factors and the associated daylight contribution

Zone	*DF (%)*	*Daylight contribution*
Bright	>6	Very large
Average	3–6	Good
Dark	1–3	Fair
Very dark	0–1	Poor

To take into account the impact of the deterioration of room surfaces, the presence of window bars on the reduction of solar radiation and the transmission of window glass, a number of correction factors should be introduced to the above equation. Thus the equation becomes:

$$DF = (DF_{SC} + DF_{EXT} + DF_{INT} \times MF) \times FR \times GL \times MG \tag{3.43}$$

where MF is the maintenance factor for room surfaces, FR is a factor related to the light loss due to window bars, GL is a correction factor due to glass transmission losses and MG is related to window maintenance. Typical values of the above-mentioned correction factors are presented in Tables 3.8–3.11.

Table 3.8 Typical values of the factor MF

	Non-industrial type of activity	*Industrial type of activity*
Non-industrial	0.9	0.7
Dirty industrial	0.9	0.6

Table 3.9 Typical values of the factor FR

Large paned windows	*FR*
All metal windows	0.8
Metal windows in wooden frames	0.75
All wood windows	0.7

Table 3.10 Typical values of the factor GL

Material	*Diffuse transmittance*	*GL*
6 mm clear glass	0.85	1
6 mm rough cast	0.8–0.85	0.95
Plastic moderately diffusing	0.75–0.8	0.9

Table 3.11 Typical values of the factor MG

Location of the building	*MG for non-industrial work*	*MG for industrial work*
Non-industrial area, vertical glazing	0.9	0.8
Industrial area, vertical glazing	0.8	0.7

The sky component DF_{SC} can be estimated using graphical tools such as the following:

- **The BRE overcast sky protractors** (Figure 3.26). These protractors consist of a set of 10. Five are for overcast sky conditions and five for uniform. There are protractors for vertical, 30°, 60°, horizontal and one protractor for windows without glazing. Each BRE protractor is composed of two parts. One part is used only with section drawings while the other is used only for plan drawings to establish the correction factor for the width of the aperture.

- **The Waldram diagram method** (Figure 3.27). In this method a series of droop lines are used to establish the view of the sky through the aperture and to estimate the DF due to sky. Its basic principle is the use of the equal area method whereby the view factor of the sky plotted on the diagram describes an area proportional to the daylight represented by the view.
- **The Pilkington sky dot method** (Figure 3.28). This method is also called the pepper dot method and is an orthographic projection technique in which the user constructs an interior perspective using the point where the calculation of illuminance is to be performed as the perspective reference point. An elevation view of the aperture is overlaid on the sky dot chart. The horizontal line of the diagram is placed on the true horizon. The sky component of DF (%) is simply the number of dots in the area of the aperture divided by 10.
- **The Bryan sky component protractors.** This set of protractors was developed for clear and overcast conditions. The protractors are used similarly to the BRE protractors but are based, in the case of clear sky, on the diffuse illuminance falling on a horizontal level from the clear sky.

The estimation of DF_{SC} in geometrically complicated scenes needs the use of a computer model. The external reflected component could be estimated by calculating the obstructed area of the sky, estimating the DF_{SC} for the obstructed part and then converting this value to DF_{EXT} by multiplying it by the ratio of the luminance of the obstructed area to the sky luminance. A simpler approach is to multiply the DF_{SC} by the average reflectance of the obstruction:

$$DF_{EXT} = DF_{SC} \times \rho$$

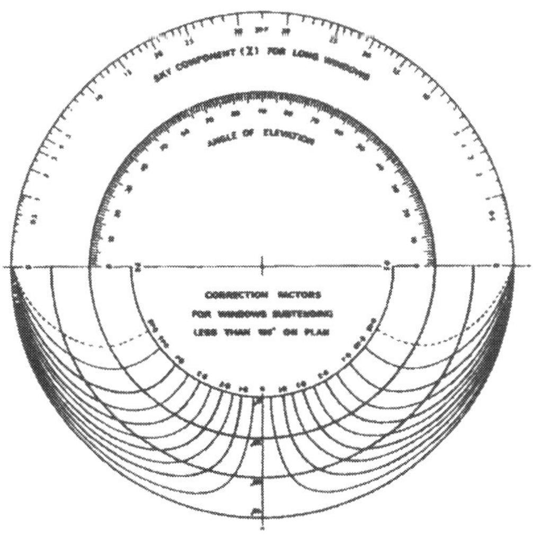

Figure 3.26 BRE protractor for vertical aperture

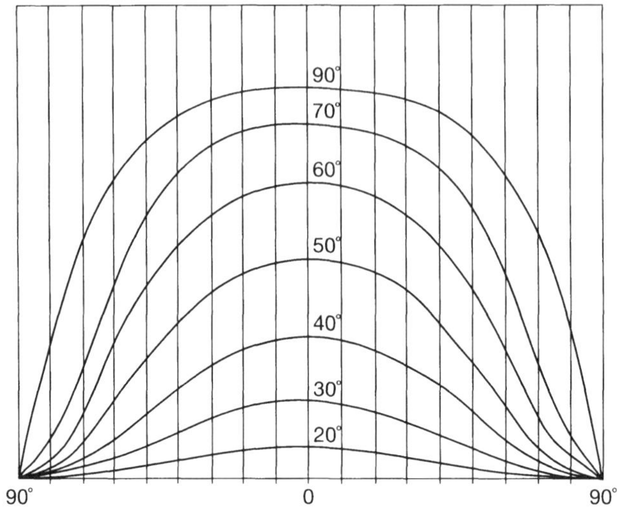

Figure 3.27 Waldram diagram

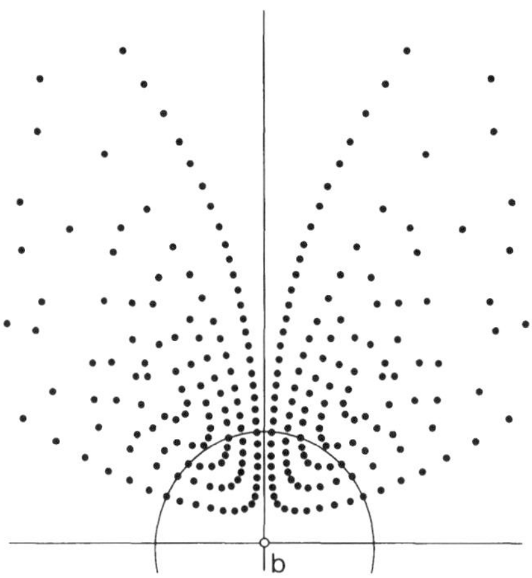

Figure 3.28 Pilkington sky dot diagram.

Under overcast conditions DF_{EXT} is much lower than DF_{SC}.

The most common approach to estimate DF_{INT} is to establish an average internal reflected component value for the entire room. This can be performed by using the split-flux method developed by BRE:

$$DF_{INT} = \frac{\tau A}{A_t(1-\rho)}(C_1\rho_u + C_2\rho_l) \tag{3.44}$$

where A_t is the sum of all the room surfaces, glazed and opaque, A is the area of the opening, and τ is the transmittance of the glazing. ρ_u and ρ_l are the average reflectance of the room, the reflectance of the upper and lower parts respectively (split along the middle of the window). The coefficients C_1 and C_2 represent the impact of external obstructions and ground. Typical values are listed in Tables 3.12 and 3.13.

Table 3.12 Values for coefficient C_1

Angle of obstruction (°)	C_l
No obstruction	39
10	35
20	31
30	25
40	20
50	14
60	10
70	7
80	5

Table 3.13 Values for coefficient C_2

Ground reflectance	C_2
0.10	5
0.20	10
0.30	15
0.40	20
0.50	25
0.60	30
0.70	35
0.80	40

The basic daylighting component is the vertical window. This permits luminous, thermal and acoustic interchange as well as natural ventilation and view. A window is characterized by its type, size, shape, position and orientation. The absolute surface of the window influences only the possibilities of ventilation and vision. Fenestration, which is expressed as the ratio between the window surface and the room area, affects the amount and distribution of light.

The performance of the window depends on the properties of the transparent element. Recently the glazing industry has made great research efforts to improve the physical and optical properties of glazing systems. Hence today there are glasses on the market with overall heat loss coefficients (U-values) from 0.4 to 5 W/m^2 K and light transmittances from 5% to 90%. There are three fundamental approaches to improving the energy performance of glazing products (two or more of these approaches may be combined):

- Alter the glazing material itself by changing its chemical composition or physical characteristics. An example of this is tinted glazing.
- Apply a coating to the glazing material surface. Reflective coatings and films were developed to reduce heat gain and glare, and more recently, low emittance and spectrally selective coatings have been developed to improve both heating and cooling season performance.
- Assemble various layers of glazing and control the properties of the spaces between the layer. These strategies include the use of two or more panes or films, low-conductance gas fills between the layers, and thermally improved edge spacers.

A new class of materials is that of translucent insulation materials (TIM). These are characterized by good optical transmission and good thermal performance. They can also replace conventional insulation. TIMs are classified into four categories:

- Multiple glazing or plastic films, with their surfaces parallel to the surface of the facade.
- Parallel slats of different materials, honeycombs and capillaries with their surface perpendicular to the surface of the facade.
- Synthetic foams, bubbles.
- Glass fibres, aerogels.

Recently, glazings with control capabilities have appeared on the market, for example electrochromic glazings, thermochromic glazings and photochromic glazings.

To evaluate the **performance of a specific glazing material**, in addition to the previously mentioned parameters such as SHGC and SC there are some other parameters:

- **Coolness factor** is the ratio of the visible transmittance to the shading coefficient. For example if the value is greater than 1, the glass is considered selective and will offer better performance in a hot climate than a similar glazing with a lower coolness factor.
- **Energy rating**. A more simplistic approach was adopted by the Canadian Window and Door Manufacturers Association. According to this approach, window energy performance is dependent on three factors: heat loss (U-value), SHGC and air infiltration. Having these essential window performance numbers, the net heat flow through a glazing can be calculated. The energy rating (ER) is a net heat flow calculation procedure and is designed to compare window products for their heating season efficiency under average winter conditions for the northern USA and Canada. This index permits a convenient standard of comparison for all glass products. ER is currently utilized by the Canadian Window and Door Manufacturers Association and is a part of Canadian

Standard CSA A440. The ER employs average weather data for winter temperatures and sunlight intensity representing several North American cities having a heating-dominated energy use pattern. The sunlight level used to calculate solar gain is an average of four building elevations. The ER is calculated using the following energy balance equation:

$$\mathrm{ER} = (72.2\mathrm{SHGC}) \times (21.9U_{\mathrm{w}}) \times (0.54L/A_{\mathrm{w}})$$

where ER is the effective energy rating (W/m^2), SHGC is the solar heat gain coefficient of total window, U_{w} is the U-value of total window (W/m^2/°C), L is the average air leakage of the window (m^3/hour) and A_{w} is the area of the window (m^2). Slightly different procedures are energy rating specific (ERS) and energy rating index (ERI). The ERS procedure determines a specific ER value for a window used in any of 13 cities throughout Canada. In addition to material options and window size, ERS is dependent on the climate of the particular location, the window to floor area ratio and the window orientation on the house. The ERI is designed to compare window products for their heating season efficiency under average winter conditions for the northern USA and Canada. The ERI employs average weather data for winter temperatures and sunlight intensity representing several North American cities having a heating-dominated energy use pattern. The sunlight level is an average of four building elevations. Air infiltration is assumed to be zero, which is a valid assumption if glass-only comparisons are being made or when comparing windows having equal air infiltration values:

$$\mathrm{ERI} = 100(0.87\mathrm{SC}I) \times U \times (T_{\mathrm{i}} - T_{\mathrm{o}})$$

where the units for ERI are BTU/hour/ft^2.
SC = shading coefficient (unit-less)
I – 22.9 BTUs/hour/ft^2 (constant for average solar intensity)
T_{i} = 70°F (constant for average inside temperature)
T_{o} = 31°F (constant for average outside temperature)
U = U-value (BTU/hour/ft^2/°F)

As the window height increases, the DF generally increases, as a greater portion of the brighter part of the sky (brightness increases towards the zenith) will be visible. Therefore the height and proportion of glazing above the working plane are more important than the actual shape or total area of a window. For this reason, tall Georgian windows provide better illumination than modern ribbon or strip windows, which are horizontally oriented and generally at a lower elevation, even for the same overall glazing area.

Decisions as to whether to increase the window height, or overall size, must consider the nature of the external view, and the proportions of the interior space, relative to the positioning of work stations. Windows that are too high may restrict the sight lines of occupants positioned close by, whereas windows which are too large may cause an unacceptable contrast between light levels next to, and far from, the external walls.

Minimum total areas of glazing for side-lit rooms are recommended as given in Table 3.14.

Table 3.14 Suggested window areas for daylighting

Depth of room from external wall (max., m)	*% of window on external wall as seen from interior (min.)*
>8	20
8–11	25
11–14	30
>14	35

Source: BS 8206: Part 2 1992 'Code of Practice for Daylighting'

The distribution of this area should ensure all parts of the room enjoy some external view and provide occupants with opportunities to focus on objects at varying distances from the working plane.

A window is one of the most fundamental components in a building. As an opening in the external enclosure, it provides a solar, thermal, acoustic and climatic relationship between the interior and exterior. When considered three dimensionally, it can manipulate light, views, sound and air to define and enhance the character and experience of any space. Variations in type, size, shape, location and orientation offer a number of design possibilities when characterising fenestration.

3.5.1 Type

A window 'type' is defined functionally by the primary purpose of the aperture, which will determine design decisions on size, shape, position, orientation and any control systems required (Table 3.15).

Table 3.15 Window design requirements

Purpose	*Design requirements*
1. Daylighting	Optimum height and size for required daylight factor
2. Natural ventilation	Positioned in the wall with respect to local wind directions and internal air currents
3. Daylighting and view	Size and sill height above floor level and exterior suited to occupant positions and external features
4. Daylighting and natural ventilation	Sizing and location must be suited to all parameters

Generally, windows will be required for light, views and ventilation so type should be distinguished by considerations of specific climatic characteristics of the building location.

3.5.2 Size

In terms of vision and ventilation, the absolute surface area (m^2) of a window is important. Size classifications are made with consideration of human scale (Table 3.16).

Table 3.16 Size classification of windows

Classification	*Surface area (m^2)*
Small	<0.5
Medium	0.5–2
Large	>2

The amount, and distribution, of light internally are affected by the ratio of total window surface to the area of the room. This is termed the **fenestration**, expressed as a percentage (Table 3.17).

Table 3.17 Classification of fenestration as % of room surface area

Classification	*%*	
Very low	<1	Risks low illumination, especially if low overcast skies, atmospheric pollution or external obstructions occur
	1–4	
Medium	4–10	
High	10–25	Problems with thermal control and glare unless controls used
Very high	>25	

Design decisions regarding the provision of one large window, or several smaller ones of the same total surface area, must recognize that while the amount of light admitted will be comparable, views, natural ventilation patterns and internal distribution of the light will be markedly different, particularly if the smaller windows are located in different walls. The nature of the activities within the space, external views and orientation of the building are important factors.

3.5.3 Shape

Window shapes are classified according to the ratio of height to width, which influences light distribution, depth of penetration, proportions of the view and natural ventilation strategies (Table 3.18).

Table 3.18 Effect of height/width ratio

Window shape	*Height/width ratio*	*Characteristics*
Horizontal	½	Illumination primarily parallel to the window wall; distribution is fairly unchanging throughout the day with little glare.
Intermediate	½–2	Enables a panoramic-type view.
Vertical	2	Illumination primarily perpendicular to the window wall; light distribution greatly varies throughout the day. Depth of daylighting increased but glare may be a problem.

As the window proportions become more vertical, exterior views may be laterally restricted but depth of field increases. Vertical configurations offer greater potential for natural ventilation strategies.

3.5.4 Location

The vertical location of a window relative to the height of the wall in which it is contained determines whether it is high, intermediate or low. Higher windows enable greater depth of penetration and distribution of daylight but sill levels must consider the human scale. Depending on the climate and strategy, natural ventilation may benefit from high opening lights.

The horizontal location relative to wall width is classed as central, lateral or corner. Light distribution from a central window may be greater, but glare is less of a problem in corner locations. The choice of location must consider occupant sight lines when seated and standing, relevant external view features, and layout of working planes/tasks requiring daylight.

3.5.5 Orientation

Windows should be oriented with respect to the sun path so that comfortable internal daylight levels are obtained, without excessive solar heat gain, glare or contrast.

The climatic, geographical and legislative (planning, fire) conditions specific to each site will determine the optimum window configurations, but the following effects generally apply:

- **South facing**: high luminous levels. High solar heat gain in winter, medium in summer. Illumination is variable throughout the day. Preferable for cold/temperate climates in the northern hemisphere (solar gain)

- **North facing**: low luminous level. Low solar heat gain, illumination is fairly constant throughout the day. Preferable for hot climates in the northern hemisphere and cold/temperate climates in the south (solar gain).
- **East and west facing**: medium luminous levels. High energy gain in summer, low in winter. East orientation has high illumination in the morning, west is high in the afternoon.

East- and west-facing windows can be considered comparable, although their maximum light levels occur at different times of the day, so any control systems (in terms of solar shading or heat gain) should be movable.

Fixed systems are more suitable for south- and north-facing windows, next to which activities requiring higher light levels should preferably be located.

3.5.6 Roof lights

When properly designed and located, **horizontal** roof lights admit up to three times more daylight per square metre of glazing than vertical windows, as the brightness of the sky increases towards the zenith. This light will be distributed more uniformly throughout the space over a day, and glare can be controlled by translucent glazing, shading devices and/or redirection of the light. Roof/ceiling apertures limit views to a portion of the sky, so visual contact with the landscape may be lost. Site-specific conditions will determine whether this is acceptable or not.

Vertical roof lights are sometimes preferred in single-storey deep plans, as horizontal glazing experiences greater solar heat gain in summer than winter, so may require integration with a natural ventilation system. Various configurations of horizontal or vertical lights are available.

Clerestory, lanterns, north lights and monitor roof windows are openings in the vertical plane which increase internal illuminance with predominantly zenithal light, usually diffused. Typically incorporated into the roof construction, natural ventilation can be provided if the apertures can be opened.

Skylights are openings in a horizontal or tilted roof that admit zenithal light into the space under the skylight, and can be opened for ventilation. Generally in standard units of up to 1 m width and length, they can be larger and with a variety of shapes. Hemispherical roofs may be partially or fully constructed with translucent materials, such as glass, polycarbonate, acrylic or fibreglass.

Some key ideas about the design of windows are the following:

- Feasibility factor should be above 0.25 in order for daylight to have the potential for significant energy savings. The feasibility factor is defined as:

  $\text{WWR} \times T \times \text{OF}$

where WWR is the window to wall ratio, T is the visible transmittance and OF is the obstruction factor

- Incorporation of envelope features such as rounded edges and splayed reveals, which usually soften light contrasts.
- Spaces with windows on two sides often have better daylighting distribution and glare is reduced.
- The depth of the room should be within 1–2.5 times the window head height for balanced daylight distribution.
- The higher the window the deeper the daylighting zone.
- Strip windows provide more uniform daylight.
- Effective aperture should be around 0.3. The effective aperture is the product of visible transmittance with WWR.

3.5.7 Design and control recommendations for windows

- The best way of avoiding overheating in hot climates is to minimize the transmission of solar gains into the building interior.
- Exterior shading is always more effective than interior shading.
- Sizing of south- or near-south-facing windows in relation to thermal mass is usually best based on passive solar heating design principles.
- East- and west-facing windows should not be too large (not more than 10% of the zone floor area) because they allow high solar gains in the morning and the afternoon during summer.
- The building should be insulated from the external gains during the day and its interior from the glare from direct sunlight without diminishing the level of diffuse daylight and view.
- A good inexpensive means of reducing heat gains in the summer and heat losses in the winter is double-glazing with a low-emissivity coating on the inner pane; this keeps the exterior glazing warm, reducing the risk of condensation; the risk of winter condensation is similarly reduced for the inner pane (room-side) in winter.
- Some shading devices offer a lot of flexibility, e.g. an external blind with insulated louvres can offer shade while allowing natural ventilation when the louvres or slats are rotated to the correct angle.
- Size-fixed shading devices only as large as required based on computational or graphical techniques.
- The following glass types are suitable for the corresponding orientations (northern hemisphere):

  SW to SE: clear glass with shading (preferably external)
  SE to NW and SE to NE: reflective or absorptive glass
  NW to NE: clear glass, not much shading (e.g. a tree).

Active control of shading devices such as shutters is beginning to attract a lot of interest in passive solar building design. It is already used in some buildings such as museums and in high locations to achieve continuous remote control of motor-operated venetian blinds and shutters (Lee and Selkowitz 1995). The motor controlling the shutter/blind may be connected to a photosensor and/or thermostat. Van Paassen describes computerized control of such a system for a passive solar test-cell (Chapter 8).

Lee and Selkowitz (1995) investigate the use of control strategies for coordinating the solar-optical properties of a dynamic building envelope system with a daylight-controlled lighting system to reduce electricity consumption and enhance perimeter zone comfort in commercial buildings. Their computer simulation results (with DOE-2) showed that (1) predictive control algorithms may significantly increase energy efficiency of systems with non-optimal solar-optical properties such as the automated venetian blinds, and (2) simpler on/off control strategies may suffice for more advanced envelope systems that have spectrally selective narrow-band electrochromic coatings.

Another recent development in dynamic building envelope systems is solar-acoustical-ventilated (SAV) windows (Bonhevi *et al.* 1994), which attempt to balance the need for passive solar gain utilization with the need for ventilation, good indoor air quality and sound insulation. SAV windows consist of a window with controlled ventilation inlets and outlets, built-in controlled venetian blinds, all connected to an expert controller which is also connected to the auxiliary heating/cooling system. Finally, a European advanced window information system (WIS) is available for determining the thermal and solar characteristics of advanced windows (Dijk and Bakker 1994).

3.6 Bibliography

Aleo F., Sciuto S. and Viedera R., (1994), 'Solar transmission measurements in outdoor conditions of non-homogeneous shading devices', *European Conference On Energy Performance and Indoor Climate in Buildings*, Lyon, France, Nov, 1074–1079.

Arulanantham M. and Kauskika N.D. (1994) 'Global radiation transmittance of transparent insulation material', *Solar Energy*, **53**: 4, 323–328.

ASHRAE (1989) *ASHRAE Handbook – Fundamentals*, Atlanta, GA, USA.

ASTM (1989) Calibrated Hot Box Test (C976), *ASTM Book of Standards*, ASTM.

Athienitis A.K. (1994), *Building Thermal Analysis*, Electronic Mathcad Book, Mathsoft, Boston, MA.

Athienitis A.K. and Haghighat F. (1992), 'A study of the effects of solar radiation on the indoor environment', *ASHRAE Transactions*, **98**, Part 1, 257–261.

Athienitis A.K. and Ramadan H. (1999) 'Numerical model for a building with transparent insulation', *Solar Energy* **67**: 1, 101–109.

Balaras C. (1996) 'Thermal comfort' in *Passive Cooling of Buildings*, James & James (Science Publishers) Ltd, London, Chapter 6.

Blanchet Y. and Girard M. (1993) 'Thermal and visual effectiveness of shading devices', *3rd European Conference on Architecture*, Florence, Italy, May, 153–156.

Bonhevi F., Trias A. and Traver X. (1994) 'Ventilation techniques integrated in the building envelope', *European Conference On Energy Performance and Indoor Climate in Buildings*, Lyon, France, Nov., 860–865

Brandemuehl M.J. and Beckman W.A. (1980) 'Transmission of diffuse radiation through CPC and flat plate collectors glazing', *Solar Energy* **24**, 511-513.

Braun P.O., Goetzberger A., Schmid J. and Stahl W. (1992) 'Transparent insulation of building facades – steps from research to commercial applications', *Solar Energy*, **49**: 5, 413–427.

BRE (1986) *Estimating daylight in buildings*, Digest 310, BRE.

Brunello Z. (1987) *Solar Properties of Fenestration, Solar and Thermal Properties of Windows: Expert Guide*, Annex XII, IEA.

Buchberg H. (1969) 'Sensitivity of room thermal response of buildings to perturbations in the climate', *Building Science*, **4**, 43–61.

Button D. and Rte B. (1992) *Glass in Building*, Pilkington Ltd., Butterworth, London.

Cane R.L.D., Hollands K.G.T., Raithy G.D. and Unny T.E., (1977) 'Free convection heat transfer across inclined honeycombs panels', *ASME Transactions*, Feb, 86-91.

CEC (1993) *Daylight in Architecture, a European reference book*, CEC, DGXII, James & James (Science Publishers) Ltd, London.

Charters W.W.S. and Peterson L.F. (1972) 'Free convection suppressing using honeycomb cellular materials', *Solar Energy*, **13**, 353–361.

Chattha J.A. and Jesch L.F. (1989) 'Solar transmission potential of transparent honeycomb insulation', *Proceedings of the ISES World Congress*, Japan.

Cogan et al. (1986) 'Optical switching in complementary electrochromic windows', *Pro. SPIE Conference*, V. 692, San Diego, CA.

Coronel J.F., Alvarez S. and Molina J.L. (1994) 'Solar-optical and thermal performance of Louver's type shading devices', *European Conference On Energy Performance and Indoor Climate in Buildings*, Lyon, France, Nov, 691–696.

Couret, G., Paule B., Scartezzini, (1994), 'Anidolic Zenithal Openings', *European Conference On Energy Performance and Indoor Climate in Buildings*, Lyon, France, Nov, 569–574

Dietrich U. (1993) 'A partly transparent shading system based on holographic optical elements', *3rd European Conference on Architecture*, Florence, Italy, May.

Dijk H. and Bakker L.G. (1994) 'Development of a European advanced windows information system', *European Conference On Energy Performance and Indoor Climate in Buildings*, Lyon, France, Nov., 920–935.

Dijk H.A.L. and Linden G.P. (1994), 'The PASSYS test method for thermal and solar characterization of building components', *European Conference On Energy Performance and Indoor Climate in Buildings*, Lyon, France, Nov., 1092–1097.

Edwards D.J., Arnold J.R. and Catton I. (1976) 'End clearance effects on rectangular-honeycomb solar collectors', *Solar Energy*, **18**, 253–257.

Elsherbiny S.M., Raithby G.D. and Hollands K.G.T. (1982), 'Heat transfer by natural convection across vertical and inclined air layers', *Journal of Heat Transfer*, **104**:1, 96–102.

Energy in Architecture – The European Passive Solar Handbook, 1992, EC Publication No. 13446, Batsford, London.

Fellands J.R. and Edwards D.K. (1978) 'Solar and infrared radiation properties of parallel plate glass honeycomb", *Journal of Energy*, **2**:5, 253–257.

Feuermann D. and Novoplansky A., 'Reversible low solar heat gain windows for energy savings', *Solar Energy* (special issue: advanced glazings), **62**:3, 169–175.

Francia G. (1961) 'Un nouveau collecteur de l'energie rayonnante solaire – theorie et verifications experimentales', *Conference des Nation Unies sur les sources nouvelles d'energie*, E/conf.35/S/71, 554–558.

Fuji T. and Imura H. (1972) 'Natural convection heat transfer from a plate with arbitrary inclination, *International Journal of Mass Transfer*, **15**, 755.

Guthrie K.I. and Charters W.W.S. (1982) 'An evaluation of a transverse slatted flat plate collectors', *Solar Energy*, **28**:2, 89-97.

Goetzbeger A., Schmid J., Wittwer V (1984), 'Transparent insulation system for passive solar energy utilization in buildings', *International Journal of Solar Energy*, **2**, 289–308.

Griffiths J. (1990) 'Thermal comfort in buildings with passive solar features: field study', *Final Report to CEC*, University of Surrey, Guildford, UK.

Hollands K.G.T. (1965), 'Honeycomb devices in flat-plate collectors', *Solar Energy*, **9**, 159.

Hollands K.G.T. and Iyankaran K. (1985) 'Proposal for a compound honeycomb collector', *Solar Energy*, **34**:4, 309–316.

Hollands K.G.T., Marshall K.N. and Wedel R.K. (1978), 'An approximate equation for predicting the solar transmittance of transparent honeycomb', *Solar Energy*, **21**, 231–236.

Hollands K.G.T., Raithby G.D. and Unny T.E. (1976*), Studies on methods of reducing heat losses from flat plate solar collectors, Phase I report*, University of Waterloo.

Holman J.P. (1986) *Heat Transfer*, McGraw-Hill Inc. New York.

Hoogendoorn C.J. (1985) *Natural Convection*, Hemisphere Publ. Corp.

Hottel H.C. (1976) 'A simple model of transmittance of direct solar radiation through clear atmospheres', *Solar Energy*, **18**.

IHVE (1971) *IHVE Guide Book 1970, Book A*, Institution of Heating and Ventilating Engineers, London.

Jesch L.F. (1993) 'Introduction', *Proceedings of the Sixth International Meeting on Transparent Insulation Technology*, Birmingham, UK, p5.

Jesch L.F., Chattha J.A. and Jankovic L. (1987), 'Transparent insulation for retrofit applications', *Proceedings of ISES World Solar Congress*, Hamburg, FRG.

Kaushika N.D. and Padma Priya R. (1991) 'Solar transmittance of honeycomb and parallel slat arrays', *Energy Conversion Management*, **32**:4, 345–351.

Kenny P. (1999) 'Design with daylight' in *FOS ToolBook*, CEC, DGXVII.

Kimura K. (1977) *Scientific Basis for Air-Conditioning*, Applied Science Publishers Ltd.

Knudsen H.N., deDear R.J., Li T.L., Puntner T.W. and Fanger P.O (1989) 'Thermal comfort in passive solar buildings', *Final Report*, CEC Research Project EN3S-0035-DK(B), Laboratory of Heating and Air Conditioning, Technical University of Denmark, May.

Kristensen P.E. (1994) 'Daylighting design of European buildings', *European Conference On Energy Performance and Indoor Climate in Buildings*, Lyon, France, Nov., 587–595.

LBL (1988), Window 3.1 Program, LBL25-686, USA.

Lee E. and Selkowitz S. (1995) 'The design and evaluation of integrated envelope and lighting control strategies for commercial buildings', *ASHRAE Transactions*, **101**, Part 1.

Lee J. *et al* (1986) 'Thermochromic materials research for optical switching', *Proc. SPIE Conference* V.692, San Diego,CA.

Littlefair P. (1995) 'Light shelves: computer assessment of daylight performance', *Lighting Research and Technology*, **27**:2, 79–91.

Liu B. and Jordan R. (1960) 'The interrelationship and characteristic distribution of direct, diffuse, and total solar radiation' *Solar Energy*, **4**,1–19.

Marshall K.N., Wedel R.K. and Damann R.E. (1976) 'Development of plastic honeycomb flat-plate solar collectors', *Report*, Lockheed Missiles and Space Company Inc, Palo Alto, California, USA, Contract No. SAN/1081-76/1.

MathSoft (2000), *Mathcad 2000*, Matsoft Inc., Boston, MA.

McMurrin J.C. and Buchberg H.B. (1971) 'Design optimization of sinusoidal glass honeycomb for flat plate solar collectors', *Journal of Solar Engineering*, **103**, 268–274.

Morris P.A. (1976) 'Radiative transfer through thin walled glass honeycomb', *ASME* paper No. 76-HT-48.

Narenda Basal, Hauser G. and Minke G. *Passive Building Design*, Elsevier Science.

Olsen L. (1982) 'Transparent insulation for thermal storage walls', *Solar Energy R & D in the European Community Series A*, D. Reidel Publishing Company, Dordrecht, **2**, 127.

PASSYS II (1994) Final Report, CEC.

Pennigton, C.W. (1964) 'Experimental analysis of solar heat gain through insulating glass with indoor shading', ASHRAE Transactions, **70**, 54–68

Perrot M. (1967) 'Les structures cellulaires antirayonament et leurs applications industrielles', *Solar Energy*, **101**, 34–40.

Platzer W.J. (1992) 'Total heat transport data for plastic honeycomb-type structures', *Solar Energy*, **49**:5, 351–358.

Platzer,W. J. (1994) 'Transparent insulation materials: A review', *SPIE Conference Proceedings* 2255, Optical Materials Technologies for Energy Efficiency and Solar Energy XIII, 616–627.

Ramadan H. (1998) 'Numerical simulation model of a test-room with transparent insulation', M.A.Sc. Thesis, Concordia University, Montreal, Canada.

Santamouris M. (1994) *PASSPORT Manual*, CEC.

Santamouris M. *et al.* (1994) *Passive Cooling of Buildings*, CIENE, University of Athens.

Sullivan H.F. (1987), VISION computer program, University of Waterloo, Canada.

Symons J.G. (1982) 'The solar transmittance of some convection suppression devices for solar energy applications: an experimental study', *Journal of Solar Energy Engineering*, **104**, 251–256.

Symons, J.G and Peck M.K. (1983) 'An overview of the CSIRO project on advanced flat solar collectors', *Proceedings of the ISES Solar World Congress*, Perth, August, 748–752.

Tabor H. (1969) 'Cellular insulation (honeycombs)', *Solar Energy*, **12**, 549–552.

Tips for daylighting, The integrated approach, LBNL-39945.

Trombe F., Robert J.F, Cabonot M. and Sesolis B, (1977) 'Concrete walls to collect and hold heat', *Solar Age* **2**, 13.

Tsangrassoulis A. (1997) *Daylight techniques in buildings, Reference Manual*, Chapter 8.

Twidell J.W, Johnstone C., Zuhdy B., Scott A. (1994), 'Strathclyde university's passive solar, low energy residences with transparent insulation', *Solar Energy*, **52**:1, 85–109.

Vandaele L. and Wouters P. (1994) *The PASSYS Services*, Summary report, European Commission, Directorate General XII for Science, Research and Development, Belgium.

Ward G.J., Rubinstein F.M. and Clear R.D. (1988) 'Ray tracing solution for diffuse interreflection', *Computer Graphics*, **22**:August.

Warres M., Selkovitz S., Morse O., Benton C. and Jewell J. (1986) 'Lighting system performance in an innovative daylight structure: an instrumental study', *Proceedings of the 2nd International Daylighting Conference*, Long Beach, CA.

Wilkinson D. (1992) 'Modelling attenuation by urban trees', PhD thesis, Manchester Metropolitan University.

CHAPTER 4

Dynamic models of heat transfer in solar buildings

4.1 Introduction

During the thermal analysis and design of a passive solar building it is necessary to evaluate heating or cooling loads and room temperature fluctuations with given weather data, such as solar radiation and ambient temperature. Moreover, it is desirable to evaluate the building response under extreme weather conditions for many design options, each time changing only a few of the building parameters, until an optimum response is obtained. Thus, it is desirable to have efficient simulation and design tools that can be used for routine passive solar analysis.

In developing methods for building thermal simulation and analysis two basic steps are involved. The first step is to develop a mathematical model describing the heat transfer processes; the thermal exchanges are usually modelled as accurately as possible, while at the same time the model complexity is kept within reasonable limits so that it can be used for practical analysis and design. The second step is to develop an appropriate method of solution for the variables of interest (usually room temperature and auxiliary energy loads) and possibly a method of analysis of the system without necessarily performing actual simulation.

The desirable degree of detail in the analysis of a building depends on the design stage. For conceptual design or the very early stages of design, a steady-state model is often satisfactory. For preliminary design more detail is required, taking into account all objectives of thermal design and the specific characteristics of the system considered.

The primary objective in the design of a passive solar building is to prevent overheating while at the same time achieving high savings in energy consumption. In direct gain systems the solar energy transmitted through south-facing windows is stored directly in the space where it is to be used, that is, in storage mass distributed in the room interior. In indirect systems the energy is stored in specially built storage elements such as a rockbed or a Trombe wall (a collector–storage wall).

In passive solar buildings the natural energy flows are the most important, and controlled auxiliary heating becomes less significant than in thin-walled conventional residential buildings in which intermittent heating is required to respond to sudden changes in the weather. The passive solar building is essentially a low-pass thermal filter that highly attenuates the high-frequency fluctuations of the weather inputs.

There exist numerous simulation and design methods for buildings. These methods are described in handbooks (ASHRAE, etc.). The work done in building energy analysis before the surge of interest in passive solar cannot in itself be considered passive solar because solar gains in the rooms were usually considered as a nuisance rather than as a benefit and heat storage in building elements was not treated accurately. Further, no emphasis was placed on the natural energy flows in the building – the most important objective was equipment sizing.

Modelling of the room interior radiant heat exchanges is more important with direct gain than with indirect gain systems and generally requires more modelling detail. In designing direct gain buildings (that is, a building with at least one direct gain room) the basic aim is to store energy in the walls during the daytime for release at night without having uncomfortable temperature swings.

A basic characteristic of passive solar building is the strong convective and conductive coupling between adjacent rooms which constitute thermal zones. This coupling is very important between south-facing direct-gain rooms receiving a significant amount of solar radiation transmitted through large windows and adjacent north-facing rooms which receive very little solar radiation. For example, heat transfer by natural convection through a doorway connecting a warm direct-gain room and a cool north-facing room can be an effective way of heating the cool room.

The design of direct-gain buildings can be separated into two phases. First, we have determination of room temperature swings on relatively clear days during the heating season (assuming no energy dumping) in order to decide how much storage mass to include so as to ensure that overheating does not occur frequently. Second, to determine the optimum amount of insulation and window area and type, the net increase of the room mean (daily or monthly) temperature above ambient temperature due to the solar gains is calculated, or auxiliary heating loads are computed until the desired energy savings are achieved.

Periodic conditions are usually assumed (explicitly or implicitly) in dynamic building thermal analysis and load calculations. Heating or cooling load, that is the auxiliary heat energy input/removal required to maintain comfort conditions, is usually calculated for a design day. The peak heating load is used to size heating equipment and the peak cooling load to size cooling equipment.

The approximations commonly introduced in mathematical models describing the building thermal behaviour can be separated into three categories:

- **Linearization of heat transfer coefficients**. Convective and radiative heat transfer are non-linear processes and the respective heat transfer coefficients are usually linearized so that the system energy balance equations can be solved by direct linear algebra techniques such as Gaussian elimination. The appropriate linear system matrix is time-invariant and its inverse can be obtained once and for all for a certain set of system parameters.

- **Spatial and/or time discretization**. The equation describing conduction is a parabolic, diffusion-type partial differential equation. Thus in using finite difference methods, a conducting medium with significant thermal capacity must be discretized into a number of regions which are modelled by lumped elements. Also, time domain discretization is required in which an appropriate time step is employed. In response factor methods only time discretization is necessary. For frequency domain analysis none of these approximations is required; in periodic models, however, the number of harmonics employed must be kept within reasonable limits.
- **Approximations for reduction in model complexity**. These approximations are usually employed in order to reduce the number of simultaneous equations to be solved or to enable the derivation of closed-form analytical solutions. They are by far the most serious and error-prone approximations and are usually done by combining radiative and convective heat transfer coefficients, assuming that many surfaces are at the same temperature, or completely ignoring certain heat exchanges.

A major aspect of the modelling process considers heat conduction in the building envelope. In most cases relating to heating or cooling load estimations, energy savings calculations and thermal comfort analysis, it is generally accepted that one-dimensional heat conduction may be assumed. Thermal bridges such as those present around corners and at the structure are generally accounted for in calculating the thermal resistance of building envelope elements. However, the thermal storage process may be adequately modelled as one dimensional.

Consider again the one-dimensional heat diffusion equation given in Chapter 2:

$$\partial T/\partial t = \alpha \partial^2 T/\partial x^2 \tag{4.1}$$

with boundary conditions
(i) at $x = 0$, $q = -k\partial T/\partial x = h_o(T_{eo} - T_1)$
(ii) at $x = L$, $q = h_i(T_2 - T_r)$
where α is thermal diffusivity, k thermal conductivity, q heat flux and h heat transfer coefficient.

One-dimensional conduction through the walls is generally assumed, as well as uniform irradiation of their surfaces by solar radiation. The above equation is usually solved with finite-difference techniques, response factor methods and frequency domain techniques (Carter 1980). The main advantage of finite difference methods is that they can accurately model the system non-linearities, but they are more cumbersome and less efficient than the other two approaches; response factor and frequency domain techniques do not require spatial discretization of wall, but they are linear techniques which utilize the superposition theorem.

In using finite difference methods, a conducting medium with significant thermal capacity must be discretized into a number of regions which are modelled by lumped elements. Also, time domain discretization is required in which an appropriate time step is employed. An energy balance at all nodes (i) may be expressed in the form:

$$C_i(\mathrm{d}T_i/\mathrm{d}t) = Q_i - B_iT_i + \Sigma_j U_{ij}(T_j - T_i) \tag{4.2}$$

where U_{ij} is the internode heat transfer conductance, C_i is the capacity of node i, and $Q_i - B_iT_i$ is the source term (due to specified temperatures Q_i and heat inputs T_i). Finite difference models may employ explicit, implicit or mixed discretization schemes.

In response factor methods only time discretization is necessary. For frequency domain analysis none of these approximations is required; in periodic models, however, the number of harmonics employed must be kept within reasonable limits.

Response factor models form the basis of large building energy simulation programs such as DOE. They usually assume that the system is linear and time-invariant and use the superposition principle. Wall response factors relate the heat flow at the surface of a wall to the history of heat flows into the building. The heat flow $q^{\text{in}}(t)$ at the inside surface of a wall may be described by the following discrete convolution equation (Kimura 1977):

$$q^{\text{in}}(t) = \sum_{n=1}^{\infty} W_n T^{\text{in}}(t - n\Delta) + M_n T^{\text{out}}(t - n\Delta) \tag{4.3}$$

where W_n and M_n are response factors for inside and outside temperature respectively and n is the number of time lags. The response factors must be calculated for the time step Δ (usually one hour) and this reduces the flexibility of the method (i.e. precalculated response factors can only be used for the same time step as used in determining the factors).

Calculation of response factors can be done using Laplace transforms, and no discretization is required. A more efficient method, the z-transform, can also be used (Stephenson and Mitalas 1971).

4.1.1 Thermal mass characteristics and implications

The modulating-filtering action of thermal mass has two effects:

- Heat gain into the room is reduced and the cooling load drops; in winter, passive solar gains are effectively stored.
- The reduction in surface temperature results in a drop of the room mean radiant temperature and the operative temperature (ASHRAE 1989), enhancing thermal

comfort in the cooling season; in the heating season solar gains are stored with a small temperature swing.

The thermal properties of major thermal storage materials and a few other materials (for comparison) are given in Table 4.1. Generally, thermal mass has high thermal capacitance but low thermal resistance. For example in 1 m^2 of concrete 10 cm thick we can store (for 1°C temperature rise)

$$Q = c\rho \text{ vol} = 840\ (\text{J/kg°C}) \times 2200\ (\text{kg/m}^3) \times 0.10\ \text{m} \times 1\ \text{m}^2$$
$$= 184\ 800\ \text{J}$$

By contrast, its thermal resistance is only 0.1/1.7 K/W, i.e. negligible.

The equation describing conduction is a parabolic, diffusion-type partial differential equation. Thus, in using finite difference methods, a conducting medium with significant thermal capacity must be discretized into a number of regions that are modelled by lumped elements. Also, time domain discretization is required in which an appropriate time step is employed. In response factor methods only time discretization is necessary. For frequency domain analysis none of these approximations are required; in periodic models, however, the number of harmonics employed must be kept within reasonable limits.

Table 4.1 Properties of thermal mass and other building materials (ASHRAE 1989, CIBSE 1988)

Material	*Mass density* (*kg/m³*)	*Thermal conductivity* (*W/m K*)	*Specific heat* (*J/kg K*)
Heavyweight concrete	2243	1.73	840
Clay tile	1121	0.57	840
Gypsum	1602	0.73	840
Gas-entrained concrete	400	0.14	1000
Water	1000	0.58	4200
plasterboard	840	0.16	950
expanded polystyrene	25	0.035	1400
Timber			
softwood	630	0.13	1360
hardwood	630	0.15	1250
plywood	530	0.14	1214
chipboard	800	0.15	1286
Common brick (full)	1922	0.727	840
Stone			
granite	2600	2.50	900
limestone	2180	1.59	720
sandstone	2000	1.30	712
marble	2500	2.00	802
Screed finish (lightweight)	1200	0.41	840

Frequency domain analysis techniques with complex variables are usually employed for steady periodic analysis of multilayered walls. They provide a convenient means for periodic analysis, in which the main parameters of interest are the magnitude and phase angle of room temperatures and heat flows.

4.2 Admittance transfer functions for walls

This section demonstrates techniques for steady-periodic thermal analysis of wall heat flow based on frequency domain techniques and the use of simple Fourier series models for outside temperature and solar radiation. Frequency domain transfer functions such as wall admittance are studied in terms of magnitude and phase lag and are then used together with Fourier series models for the weather variables to determine steady-periodic thermal response of walls. The technique is applied for passive solar analysis and design.

Significant insight into wall dynamic thermal behaviour may be obtained by studying the wall's admittance transfer functions (magnitude and phase angle) as a function of frequency, thermal properties, and geometry.

Figure 4.1 shows conceptually how wall response to weather inputs (e.g. $T\sin(\omega t)$) may be obtained for one harmonic, and the time lag between the input and output waves. For inputs with more than one harmonic, the total response may be obtained by superposition of the response harmonics.

The thermal admittance of a wall is a transfer function parameter useful for analysis of the effects on room temperature of cyclic variations in weather variables such as solar radiation, outside temperature and dynamic heat flows under steady periodic conditions.

There are two transfer functions of primary interest, namely the self-admittance Y_s relating the effect of a heat source at one surface to the temperature of that surface, and the transfer admittance Y_t relating the effect of an outside temperature variation to the resulting heat flow at the inside surface.

These two transfer functions are determined as demonstrated in the following model (Athienitis *et al.* 1986). The wall in Figure 4.2 consists of insulation and thermally non-massive layers (low thermal capacity) with conductance value U per unit area, and a thermally massive layer of thickness L.

The Norton equivalent network for a wall with a specified temperature on one side (such as basement temperature or sol-air temperature) is obtained from the cascade form of the wall equations which relates temperature and heat flow at one surface to those at the other surface. The cascade form of the equations is derived by first taking the Laplace transformation of the one-dimensional heat diffusion equation to obtain an ordinary differential equation, which can then be solved to relate heat flux and temperature at one surface of a one-dimensional medium to those at the other surface, without discretizing.The cascade equations relating temperature T_1 and heat flux q'_1 at one surface of a slab to those (q_2 and T_2) at its other surface are given by (Kimura 1977):

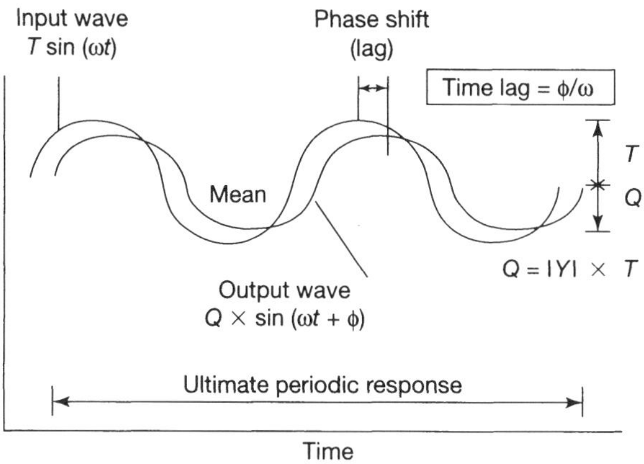

Figure 4.1 Schematic of temperature and heat flow waves (Y = admittance transfer function with magnitude $|Y|$ and phase angle φ)

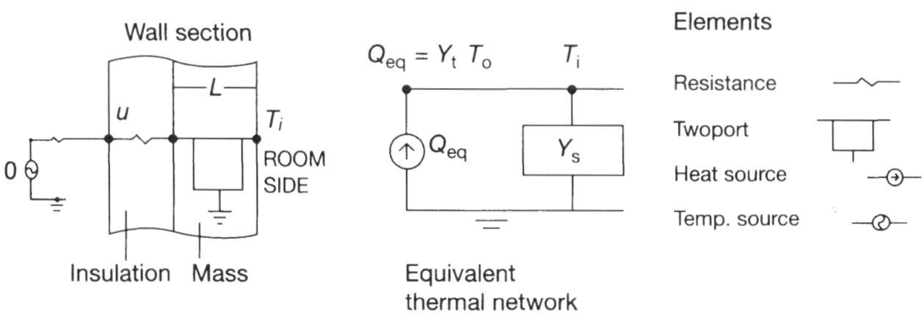

Figure 4.2 Exterior wall with massive interior layer, and equivalent thermal network

$$\begin{bmatrix} T_1 \\ q'_1 \end{bmatrix} = \begin{bmatrix} D & B \\ C & D \end{bmatrix} \cdot \begin{bmatrix} T_2 \\ -q'_2 \end{bmatrix} \tag{4.4}$$

where $D = \cosh(\gamma x)$, $B = \sinh(\gamma x)/(k\gamma)$ and $C = (k\gamma)\sinh(\gamma x)$, and q' is positive into the slab.

The quantity k is the thermal conductivity, x is thickness, and γ is equal to $(s/\alpha)^{1/2}$, s being the Laplace transform variable and α being the thermal diffusivity. For admittance calculations s is set equal to $s = j\omega$ where $j = \sqrt{-1}$ and $\omega = 2\pi/P$. For diurnal analysis, period $P = 86\ 400$ s.

The cascade matrix for a multilayered wall is obtained by multiplying the cascade matrices for consecutive layers. Usually the temperatures of interest are either the inside or the outside temperature. In this way, wall interior temperatures of no interest are easily eliminated. A linear subnetwork connected to a network at only two terminals (a port) can be represented by its Norton equivalent, consisting of a heat source and an admittance connected in parallel between the terminals (Athienitis *et al.* 1986).

The admittance is the subnetwork equivalent admittance as seen from the connection port (the two terminals), and the heat source is the short-circuited heat flow at the port. Consider for example the wall in Figure 4.2, assumed to be made up of an inner layer of storage mass of uniform thermal properties and a massless insulation layer, also of uniform thermal properties. The region behind the thermal mass may be represented by equivalent conductance U in series with the outside temperature T_o (for exterior walls the sol-air temperature T_{eo}). The conductance U combines the insulation resistance and a film coefficient.

The determination of Y_s (called the wall self-admittance), and the equivalent source Q_{sc} produced by the transformation, will now be explained. A first step obtains the total cascade matrix by multiplying the cascade matrix for the storage mass by that for u (note: $u = U/A$):

$$\begin{bmatrix} T_S \\ q_S \end{bmatrix} = \begin{bmatrix} D\,B \\ C\,D \end{bmatrix} \cdot \begin{bmatrix} 1 & 1/u \\ 0 & 1 \end{bmatrix} \cdot \begin{bmatrix} T_O \\ -q_O \end{bmatrix} \tag{4.5}$$

After multiplying, we temporarily set $T_s = 0$, to get the Norton equivalent source as

$$Q_{eq} = -Y_t T_o$$

where the transfer admittance Y_t is given by

$$Y_t = -A/[(A \cosh(\gamma x)/U) + \sinh(\gamma x)/(k\gamma)] \tag{4.6}$$

The transfer admittance has been multiplied by the area A to obtain its total value. To obtain Y_s, we temporarily set $T_o = 0$ and obtain the admittance as seen from the interior surface, yielding (after multiplying by A):

$$Y_s = A[(U_o/A) + k\gamma \tanh(\gamma x)]/[(U_o/k\gamma A)\tanh(\gamma x) + 1] \tag{4.7}$$

If there is no mass then we obtain the simple equality $Y_s = -Y_t = U$. A similar result is obtained for windows in eliminating all nodes exterior to the inner glazing. An important result is obtained for an infinitely thick wall or a wall with no heat loss at the back (adiabatic surface, or high amount of insulation); in this case Y_s is given by

$$Y_s = Ak\gamma \tanh(\gamma x) \tag{4.8}$$

Thick walls have admittance that is close to that given by the above equation. When the penetration depth, given by

$$d = \sqrt{[2k/(c\rho\omega)]} = \sqrt{(2\alpha/\omega)} \qquad (4.9)$$

is significantly less than the wall thickness then the wall behaves like a semi-infinite solid.

The magnitude, phase angle (and time lag/lead) of a transfer function such as Y_s and Y_t are computed by means of complex variables.

Analysis: substantial insight into wall and building thermal behaviour may be gained by studying the magnitude and phase angle of the important transfer functions such as Y_s and Y_t (Athienitis 1994). The time lead d_s of Y_s is the time difference between the peak of a sinusoidal input function, such as solar radiation in the case of the room interior surface, and the resulting peak of the interior surface temperature T_i. Now, we consider the variation of wall thermal admittance with thermal mass thickness L for the fundamental frequency (one cycle per day, $n = 1$) for unit wall area. Note that the diurnal ($n = 1$) frequency is important in the analysis of variables with a dominant diurnal harmonic such as solar radiation. High frequencies are important in analysing the effect of varying heat inputs such as those due to on/off cycling of a furnace.

A wall with an interior massive layer of concrete is considered and exterior insulating layers with insignificant thermal capacity and thermal resistance equal to 4 RSI. The concrete has a specific heat capacity of 800 J/kg°C, a density of 2200 kg/m^3 and a thermal conductivity of 1.7 W/m K. The results presented below are specific to this concrete, but they generally indicate the expected trends for concrete/brick/masonry type materials. Note that the thermal conductivity of these materials increases with moisture content and density.

Figure 4.3 shows an extremely important result in steady-periodic analysis of building thermal response – the fact that there is a wall thermal mass thickness that will most reduce room temperature fluctuations – in this case for $L = 0.2$ m, corresponding to the maximum admittance. Therefore, this is the optimum thermal mass thickness for passive solar design.

As indicated in Figure 4.4, the magnitude of wall admittance (for mass thickness of 20 cm) increases with frequency (decreases with period). Thus, the inside room temperature fluctuations are smaller for high-frequency fluctuations in internal heat gains. For harmonic numbers higher than about 8, that is periods less than 3 h, the wall behaves like an infinitely thick wall; in this case the phase angle is 45°.

The variation of transfer admittance Y_t with mass thickness is depicted in Figure 4.5. In this case, the magnitude of Y_t decreases with thickness and therefore the fluctuations of the sol-air temperature are significantly modulated as they are transmitted to the room interior. This is a well-known phenomenon, efficiently employed in traditional architecture in adobe buildings. The time lag of the heat gains transmitted ($q = Y_t \times T_{eo}$) into the interior is the time lag of Y_t. This time lag increases to about 7.5 h for a mass thickness of 30 cm.

The time delay in the transmission of a heat wave through a wall is another positive effect of thermal mass in addition to the attenuation of the temperature

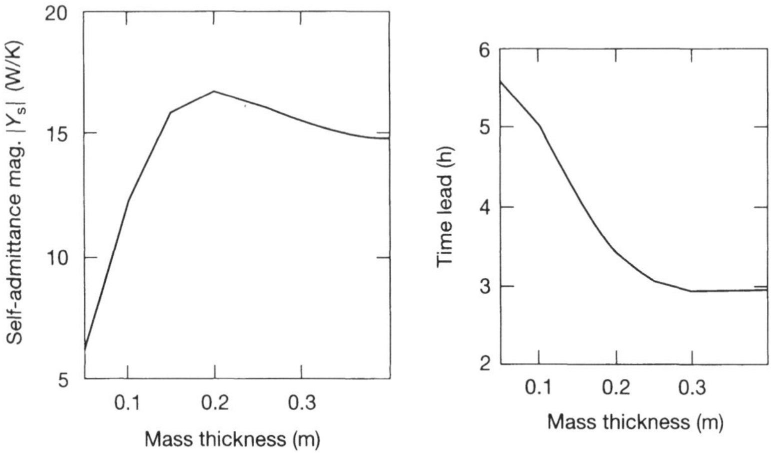

Figure 4.3 Variation of the self-admittance and its time lead with mass thickness

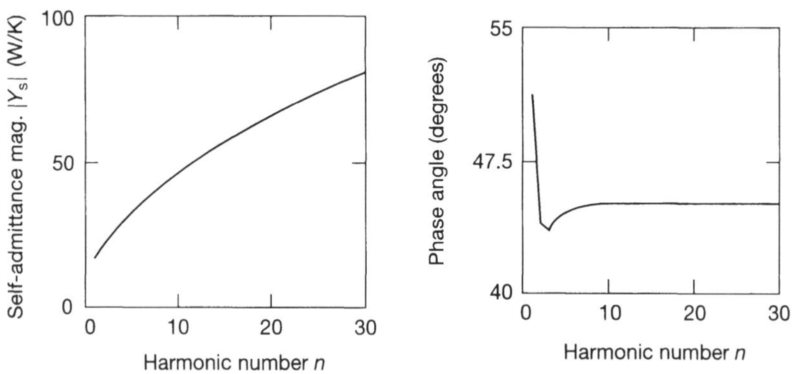

Figure 4.4 Variation of the self-admittance and its phase angle with harmonic number *n*

swings. Thus, the peak heat gains through the structure coincide with cooler outside conditions when natural ventilation may be employed to reduce total instantaneous cooling loads.

Figure 4.6 shows that the magnitude of the transfer admittance decreases relatively quickly with increasing harmonic number n and decreasing period. Thus, the heat gain fluctuations transmitted into the room as a result of sol-air temperature fluctuations are significantly reduced at high frequencies. For example, a temperature swing of 2°C in sol-air temperature over a period of about 2 h will result in a heat gain swing of about

$$0.1 \text{ W/°C} \times 2\text{°C} = 0.02 \text{ W (per m}^2 \text{ of wall)}$$

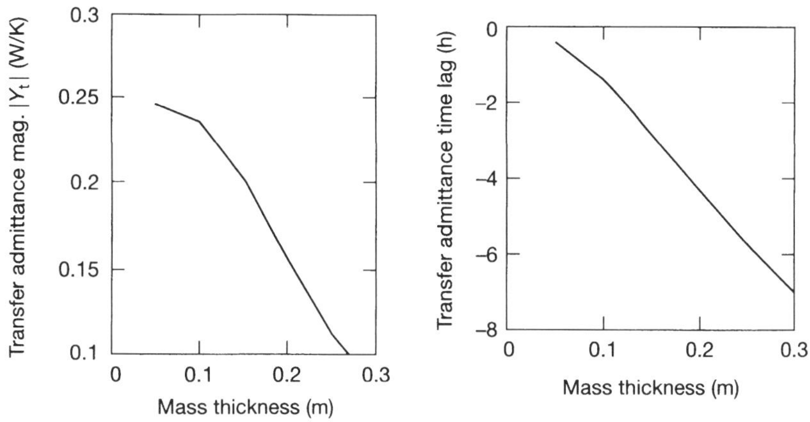

Figure 4.5 Variation of the transfer admittance and its time lag with mass thickness

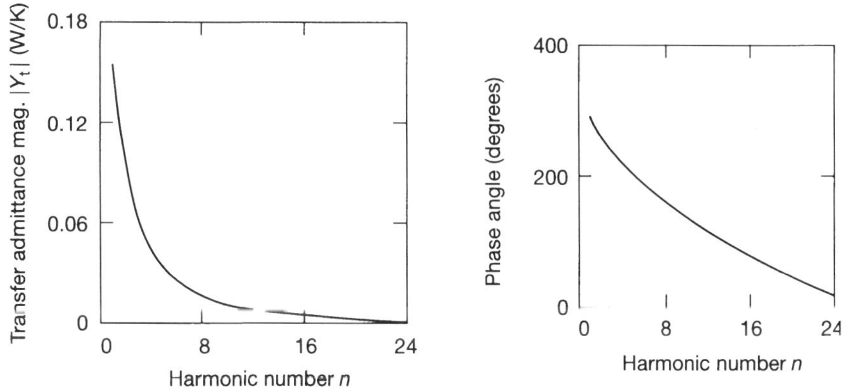

Figure 4.6 Variation of the transfer admittance magnitude with harmonic number *n*

4.3 Building transient response analysis

Transient thermal analysis of walls or rooms may be performed with the following objectives:

- Peak heating/cooling load calculations
- Calculation of dynamic temperature variation within walls, including solar effects, room temperature swings and condensation on wall interior surfaces; two-dimensional steady-state temperature profiles in walls (e.g. for investigation of thermal bridge effects).

For a multilayered wall, an energy balance is applied at each node at regular time intervals to obtain the temperature of the nodes as a function of time. These equations may be solved with the implicit method as a set of simultaneous equations or with the explicit method in which we march forward in time from a set of initial conditions.

We will consider the **finite difference thermal network approach**. In this approach, each wall layer is discretized (divided) into a number of sublayers (regions). Each region is represented by a **node** and is assumed to be isothermal. Each node (i) has a thermal capacitance (C_i) associated with it and resistances connecting it to adjacent nodes.

Wall transient thermal response analysis with finite difference techniques may generally provide a more accurate estimation of temperatures and heat flows owing to the capability to model non-linear effects such as convection and radiation. One disadvantage is that the initial conditions are usually unknown. Thus, the simulation is repeated until a steady periodic response is obtained.

The explicit finite difference method is particularly suitable for modelling of non-linear heat diffusion problems such as the present case of heat transfer through a wall. It can easily accommodate non-linear heat transfer coefficients and control actions. Figure 4.7 shows the thermal network for a wall with transparent insulation. The wall consists of an exterior layer of transparent insulation, an air cavity, and a thermal storage layer of concrete.

In the transient one-dimensional finite difference method we represent each wall layer by one or more sublayers (regions). Each region is represented by a node

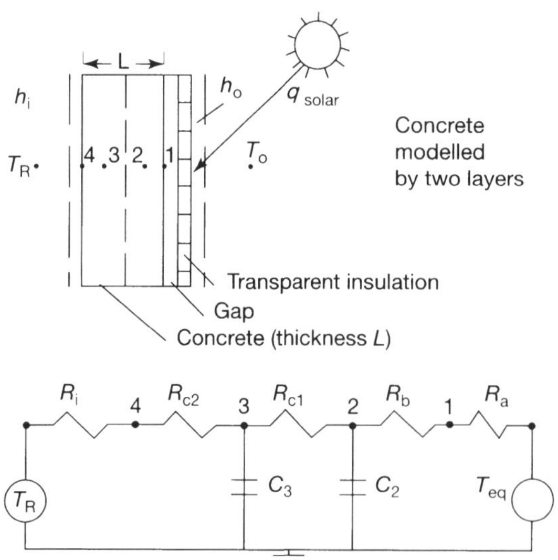

Figure 4.7 Wall with transparent insulation, and its thermal network

with a thermal capacitance C connected to two thermal resistances, each equal to half the R-value of the layer, forming a T-section as shown in Figure 4.7 for the concrete layer. Thus, the finite difference thermal network model for the wall consists of two capacitances for the concrete thermal capacity and interconnecting thermal resistances. The energy balance for the thermal network is:

$$T_{\mathrm{i,p+1}} = \frac{\Delta t}{C_{\mathrm{i}}} \times \left(q_{\mathrm{i}} + \sum \frac{T_{\mathrm{j,p}} - T_{\mathrm{i,p}}}{R_{\mathrm{i,j}}} \right) + T_{\mathrm{i,p}} \tag{4.10}$$

Subscript i indicates the node for which the energy balance is written and j all nodes connected to node i, while p is the time interval; q_i represents a heat source at node i such as solar radiation.

The time step was selected based on the following condition for numerical stability:

$$\Delta t \le \min \left(C_i \sum_j \frac{1}{R_{i,j}} \right) \qquad \text{for all nodes } i \tag{4.11}$$

We show below the specific calculations for the above model in Mathcad (Athienitis 1994) and the resulting temperature variation of a south-facing wall for two winter days with high solar gains. For summer simulations a shutter is assumed to be closed in the cavity reflecting outwards 90% of transmitted solar gains. (See also Figure 4.8.)

$L = 0.20\mathrm{m}$ $\quad A = 1\ \mathrm{m}^2$ concrete thickness and face area (perpendicular to heat flow)

Concrete properties: $C = 800\,\mathrm{J/kg°C}$ specific heat $\quad k = 1.7\,\mathrm{W/m°C}$ conductivity $\quad p = 2200\,\mathrm{kg/m^3}$ density

Film coefficients: $h_{\mathrm{i}} = 10\,\mathrm{W/m^2\,°C}$ $\quad h_{\mathrm{o}} = 20\,\mathrm{W/m^2\,°C}$

$\tau\alpha = 0.7$ effective transmittance – absorptance of transparent insulation system and concrete face

$R_{\mathrm{ins}} = 0.5\,\mathrm{m^2\,°C/W}$ $\quad R_{\mathrm{gap}} = 0.3\,\mathrm{m^2\,°C/W}$ transparent insulation and gap resistances

$R_{\mathrm{a}} = \dfrac{R_{\mathrm{ins}} + R_{\mathrm{gap}} + \dfrac{1}{h_{\mathrm{o}}}}{A}$ $\quad R_{\mathrm{a}} = 0.85\,\mathrm{°C/W}$ resistance from outside to concrete face

$R_{\mathrm{c}} = \dfrac{L}{k\,A}$ total concrete resistance $\quad R_{\mathrm{b}} = \dfrac{R_{\mathrm{c}}}{4}$ $\quad R_{\mathrm{b}} = 0.029\,\mathrm{°C/W}$

$R_{\mathrm{c1}} = \dfrac{R_{\mathrm{c}}}{2}$ $\quad R_{\mathrm{c1}} = 0.059\,\mathrm{°C/W}$ $\quad R_{\mathrm{c2}} = \dfrac{R_{\mathrm{c}}}{4}$ $\quad R_{\mathrm{i}} - \dfrac{1}{A\,h_{\mathrm{i}}}$ $\quad R_{\mathrm{i}} = 0.1\,\mathrm{°C/W}$

$C2 = \rho c \dfrac{L}{2} A$ $\quad C2 = 1.76 \times 10^5\,\mathrm{J/°C}$ $\quad C3 = C2$ thermal capacitance for each layer

Stability test

$$\mathrm{TS} = \begin{bmatrix} \dfrac{C2}{\dfrac{1}{R_a + R_b} + \dfrac{1}{R_{c1}}} & \dfrac{C3}{\dfrac{1}{R_{c1}} + \dfrac{1}{R_{c2} + R_i}} \end{bmatrix}$$

The time step Δt should be lower than the minimum of the two values in the vector TS

$\Delta t_{\text{critical}} = \min(\mathrm{TS})$ $\quad \Delta t_{\text{critical}} = 7.118 \times 10^3\ \mathrm{s}$ $\quad \Delta t = 1800\,\mathrm{s}$

$i = 0., 1..96$ number of time steps

Assume $w = 2 \cdot \dfrac{\pi}{86400} \cdot \mathrm{rad/s}$ frequency based on period of one day

$T_o(t) = \left[5\cos\left(wt + 3\dfrac{\pi}{4} \right) - 10 \right]°\mathrm{C}$ outside temperature

incident solar radiation modelled as half-sinusoid

$f(t) = 500\cos[w(t - 43200\,\mathrm{s})]\mathrm{W}$ $\quad q_{\text{solar}}(t) = \mathrm{if}(f(t) > 0\mathrm{W}, f(t), 0\mathrm{W})$

$T_{eq}(t) = T_o(t) + q_{\text{solar}}(t)\tau\alpha\, R_a$ equivalent 'sol - air' temperature at node 1 (concrete surface)

$T_R = 22°\mathrm{C}$ room temperature

Initial estimates of temperatures: $\begin{bmatrix} T2_0 \\ T3_0 \end{bmatrix} = \begin{bmatrix} 10 \\ 10 \end{bmatrix}°\mathrm{C}$

Simulation for nodes with capacitances

$$\begin{bmatrix} T2_i + 1 \\ T3_i + 1 \end{bmatrix} = \begin{bmatrix} \dfrac{\Delta t}{C2}\left(\dfrac{T_{eq}(i\Delta t) - T2_i}{R_a + R_b} + \dfrac{T3_i - T2_i}{R_{c1}} \right) + T2_i \\ \dfrac{\Delta t}{C3}\left(\dfrac{T2_i - T3_i}{R_{c1}} + \dfrac{T_R - T3_i}{R_i + R_{c2}} \right) + T3_i \end{bmatrix}$$

Finite difference simulation (i is present time and i+1 next time step)

Now we can also calculate intermediate temperatures

$$T4_i = T_R + R_i \frac{T3_i - T_R}{R_{c2} + R_i} \qquad T1_i = T2_i + R_b \frac{T_{eq}(i\Delta t) - T2_i}{R_b + R_a}$$

4.4 Ground coupling, basement heat transfer

A building is usually in direct contact with the ground either through a ground-level floor or through the floor and walls of a basement. The ground temperature does not vary as much as the ambient temperature because the ground acts as a heat sink, modulating the daily and seasonal temperature variations. Thus, the ground may be a source of heat (or reduced heat losses) in the winter while being a source of coolness in the summer.

The ground may be modelled as a semi-infinite medium and the ambient temperature may be modelled as sinusoidal (Mihalakakou *et al.* 1994, Tzaferis *et al.* 1992). The heat transfer problem is therefore:

$$\partial T/\partial t = \alpha \partial^2 T/\partial z^2 \tag{4.12}$$

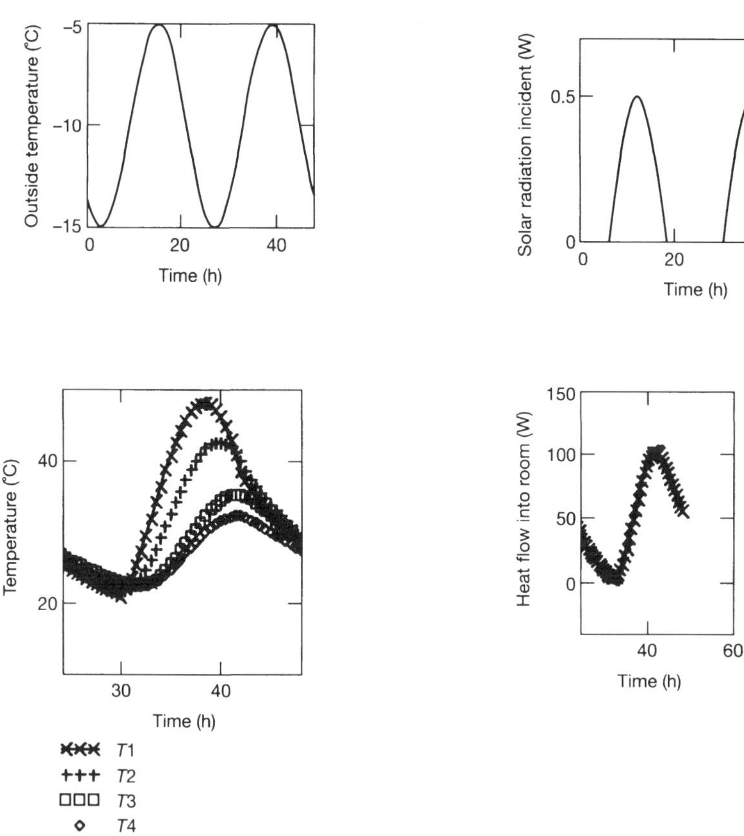

Figure 4.8 Weather conditions and temperature variation under winter conditions for wall with transparent insulation

with boundary conditions
(i) at depth $z = 0$ the ambient temperature is

$$T(t) = T_m + \Delta T \cos(\omega t)$$

with frequency $\omega = 2\pi/P$ where P is period, T_m is the mean ambient temperature and ΔT the amplitude.
(ii) at infinite depth, temperature is constant

The solution to the above equation is

$$T(t,z) = T_m + \Delta T e^{-z/d} \cos(\omega t - z/d)$$

where $d = \sqrt{(2\alpha/\omega)}$ is the penetration depth.

At a depth of d the temperature swing is about one-third of ΔT while at depths of about $5d$ the temperature variations are almost completely damped out. Figure

4.9 shows the penetration depth as a function of thermal diffusivity for the diurnal frequency ($P = 86\,400$ s) and for the yearly frequency. The thermal diffusivity of limestone soil is about 0.58×10^{-6} m^2/s, for sandstone 1.06–1.26×10^{-6} m^2/s and for marble 1.39×10^{-6} m^2/s.

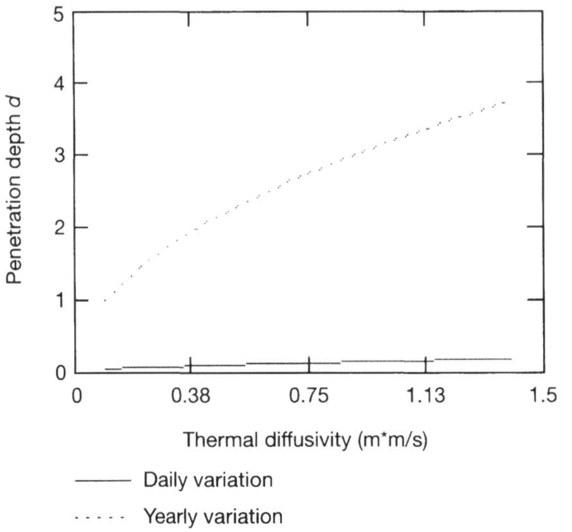

Figure 4.9 Penetration depth d as a function of thermal diffusivity

Therefore, it may be concluded that at a depth of about 1 m the diurnal temperature variation of the soil is insignificant. This is the principle on which ground cooling is based. Earth-to-air heat exchangers are employed in either open-loop or closed-loop circuits (Tzaferis *et al.* 1992). In an open-loop system air is drawn by fan from the ambient through a buried pipe into the building. During summer the air is cooled since the ground is colder than the air; a temperature differential of at least 5–6 K between ambient and the ground around the pipe is required for effective operation (Santamouris 1994). In a closed-loop system both inlet and outlet are inside the building.

For homogeneous soils with constant thermal diffusivity, the daily temperature of the ground at depth z may be represented by (Santamouris 1994):

$$T(z,t) = T_m - A_s \exp(-z\sqrt{}(\pi/365\alpha))\cos\{(2\pi/365)[t - t_o - z\sqrt{}(3/4\pi)]\} \qquad (4.13)$$

where

T_m = average annual temperature

A_s = amplitude of surface temperature variation

t = time, days, t_o = phase constant (days)

Floors in contact with the ground lose or gain heat through cyclic paths going from the floor to the surface of the ground. CIBSE (1988) recommends the following relationship for the U-value of uninsulated concrete floors (for heat transfer from/to outside air) with four edges exposed:

$$U = \frac{4ke^{b/2L}}{b\pi} \quad \tanh^{-1}\left(\frac{b}{b+w}\right) \tag{4.14}$$

where
k = soil thermal conductivity, W/m K (about 0.7–2.1)
b = breadth (smaller dimension) of floor, m
w = thickness of surrounding wall, m (about 0.3 m)
L = length of floor, m

For example, for a room with dimensions 4 × 6 m: let k = 0.7 W/m K, L = 6 m, b = 4 m. This gives U = 0.516 W/K m^2.

For floors with two parallel edges exposed use equation 4.14 for the appropriate breadth but set L equal to infinity (i.e. $\exp(b/2L) = 1$). For floors with two edges exposed the U-value is the same as for floors of twice the length and breadth with four edges exposed. For a single edge exposed U is the same as for two parallel edges but of twice the breadth. If insulation is used, add the thermal resistance of insulation to that of the floor. The U-value must be reduced when edge insulation is employed (see Table 4.2).

Table 4.2 Percentage reduction in floor U-value for insulation up to indicated depth (m) (CIBSE 1988)

Floor dimensions (m)	*% reduction for insulation up to 0.5 m*	*% reduction for insulation up to 1.0 m*
10 × 10	14	22
6 × 6	15	25
4 × 4	18	28

Basement heat transfer may be evaluated in detail numerically using finite difference techniques and in particular two-dimensional models (three-dimensional at corners).

4.5 Bibliography

ASHRAE (1989) *ASHRAE Handbook – Fundamentals*, Atlanta, GA, USA.

Athienitis A.K. (1994) *Building Thermal Analysis*, Electronic Book Mathcad, MathSoft, Boston, USA.

Athienitis A.K., Sullivan H.F. and Hollands K.G.T. (1986) 'Analytical model, sensitivity analysis, and temperature swings in direct gain rooms', *Solar Energy*, **36**, 303–312.

Carslaw H.S. and Jaeger J.C. (1959) *Conduction of Heat in Solids*, 2nd edn, Oxford University Press, London.

Carter C. (1980) 'Solving the heat transfer network equations in passive solar simulations', Proc. ASES 5th Passive Solar Conf., Amherst, MA, October, pp. 238–242.

CIBSE (1988) *CIBSE Guide*, Volume A, CIBSE, London.

Cook J. (ed.) (1989) *Passive Cooling*, MIT Press, Cambridge, MA.

Kimura K. (1977) *Scientific Basis for Air-conditioning*, Applied Science Publishers, Ltd.

Mihalakakou G., Santamouris M. and Asimakopoulos D. (1994) 'Modelling the thermal performance of earth-to-air heat exchangers', *Solar Energy*, **53**:3, 301–305.

Santamouris M. *et al.* (1994) *Passive Cooling of Buildings*, CIENE, University of Athens.

Stephenson G.D. and Mitalas G.P. (1971) 'Calculations of heat conduction transfer functions for multi-layer slabs', *ASHRAE Transactions*, **77**:2, 117–126.

Tzaferis A., Liparakis D., Santamouris M. and Argyriou A. (1992) 'Analysis and accuracy of eight models to predict performance of earth-to-air heat exchangers', *Energy and Buildings*, **18**, 35–43.

CHAPTER 5

The passive response of buildings and its use in design

5.1 Introduction

The passive response of a building is primarily reflected in its temperature variation without operation of the heating or cooling system. This response helps us assess the effectiveness of the building passive systems. In the case of heating analysis, it is useful to compare the response on a cold sunny day with that on a cold cloudy day. In both cases, the outdoor temperature profile should be the same in order to assess the solar savings by comparing the mean room temperatures in the two cases. This simple approach was introduced in Section 1.4.

However, such a simple approach does not consider the significant issue of thermal comfort in passive solar buildings. In fact, unless adequate thermal storage mass is present, the possibility of room overheating is very high for well-insulated buildings even during winter months when the exterior temperature may fall below 0°C. In this chapter a simple analytical model is presented for determining the temperature swings in direct-gain rooms for relatively clear days. It is primarily intended for study of the relative effects of fundamental design variables such as storage mass thickness and area on room temperature fluctuations.

It is useful to develop such a special-purpose simple model for studying temperature swings in a direct-gain room for the following reasons:

- The relative effects on room temperature swings of the storage mass thickness and area, the amount of insulation and the daylength may be studied by modelling the room interior as consisting of a few isothermal surfaces. Thus, an analytical solution may be obtained including only a few fundamental variables, which permits an efficient and basic sensitivity analysis.
- An analytical solution with only a few inputs permits efficient application of the superposition principle for the study of the relative importance of various weather inputs and their harmonic components.

There are several simple periodic models for determining temperature swings. However, they often employ simple heat transfer modelling assumptions, such as an isothermal (single-node) storage mass (e.g. Kirkpatrick and Winn 1984) which may introduce substantial errors. One well-known simplified design method is the diurnal heat capacity (DHC) method (Balcomb *et al.* 1980). Based on a distributed (non-isothermal) model for the storage mass, it determines the swings in room air temperature for typical clear days. Like other simplified methods, the DHC method does not incorporate the effect of change in daylength, an important parameter, as this chapter will show. Also, it only finds the room air temperature swings; however, thermal comfort is also affected by the room surface temperatures, so a weighted average of air temperature and surface temperatures is to be preferred. This chapter presents a model and a method which are an attempt to eliminate the above shortcomings.

5.2 Simple direct-gain room model and solution

5.2.1 The model

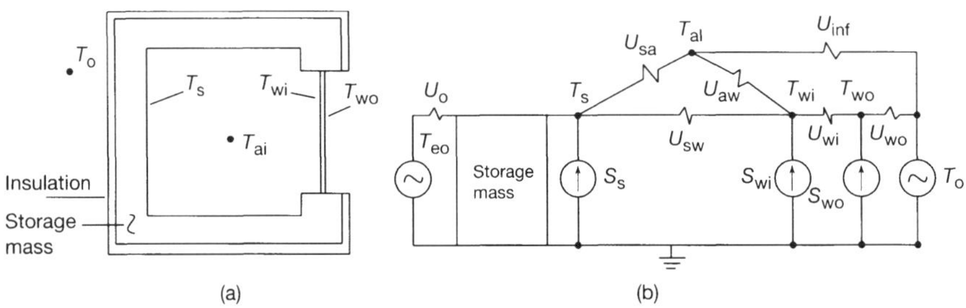

Figure 5.1 (a) Direct-gain room with uniformly distributed thermal storage mass; (b) thermal network for room with wall model by distributed RC element

The model is shown schematically in Figure 5.1(a). The walls (all of which are external) are assumed to be made up of an inner lining of storage mass material of uniform thermal transport properties, and an outer massless insulation, also of uniform thermal transport properties. Walls with storage mass are assumed to be at the same surface temperature and they are thus treated as one exterior wall, which is modelled as a two-port distributed element (Carslaw and Jaeger 1959) as described in Section 4.2; its internal surface is thus assumed to be uniformly irradiated by solar radiation. The thermal mass of the window glazings is assumed to be negligible. The important temperatures are shown in Figure 5.1(a): the storage mass internal surface temperature T_s, the (double-glazed) window internal glazing temperature T_{wi}, the room air temperature T_{ai}, and the outdoor air temperature T_o.

The assumed thermal network, shown in Figure 5.1(b), contains several conductances. The equivalent conductance U_{inf} due to infiltration heat loss links T_{ai} and T_o. Convective conductances U_{sa} and U_{aw} link T_{ai} with T_s and T_{ai} with T_{wi}

respectively, while radiative conductances U_{sw} links T_s and T_{wi}. Combined radiative and convective conductances U_{wi} and U_{wo} thermally link T_{wi} with T_{wo} and T_o respectively. The components outside the thermal mass have been replaced in Figure 5.1(b) by an equivalent conductance U_o in series with the sol-air temperature T_{eo} associated with the external wall surface. The conductance U_o combines the insulation resistance R_{ins} and the external surface radiative-convective conductance (equal to $A_o h_o$). The equations for the conductances (time-invariant) are given in equation 5.1. While the model as just described assumes a uniformly distributed storage mass, by a small extension it can model a situation where some of the walls (normally the ceiling) are non-massive, by modelling them as conductances in parallel with U_{inf}. Therefore, U_{inf} is replaced by $U_{inf} + U_{non\text{-}massive}$.

The network also contains several sources. The major source S_s is the solar radiation transmitted through the window and absorbed by the storage mass. Sources S_{wi} and S_{wo} are the rates at which solar radiation is absorbed (uniformly) by the inner and outer glazings respectively.

5.2.2 Transformation and frequency domain solution

A delta to star transformation (a three-terminal transformation; see Roe 1966) permits the network of Figure 5.1(b) to be reduced to that of Figure 5.2(a). This transformation has permitted a 'natural' representative room temperature T_{ei} to be obtained at an important node; T_{ei} is a weighted average of T_a, T_{wi} and T_s:

$$T_{ei} = (U_{ew}T_{wi} + U_{ea}T_{ai} + U_{es}T_s)/(U_{ew} + U_{ea} + U_{es}) \tag{5.1}$$

where

$$U_{ew} = U_{sw}(1 + U_{aw}/U_{sa}) + U_{aw} \tag{5.2a}$$

$$U_{ea} = U_{aw}(1 + U_{sa}/U_{sw}) + U_{sa} \tag{5.2b}$$

$$U_{es} = U_{sw}(1 + U_{sa}/U_{aw}) + U_{sa} \tag{5.2c}$$

The temperature T_{ei} can serve as a representative 'sensed' temperature because it should closely follow the effective temperature that would be felt by an occupant who senses, through radiant exchange, the temperature of the room surfaces, and through convective exchange the room air temperature.

More simplifications are applied to obtain the simplified network in Figure 5.2(b). The first transforms the minor sources S_{wi} and S_{wo} to the equivalent source S_{ei} given by

$$S_{ei} = U_{ew}[S_{wi}(h_i + h_o) + S_{wo}h_i]/[h_i h_o A_w + (h_i + h_o)U_{ew}] \tag{5.3}$$

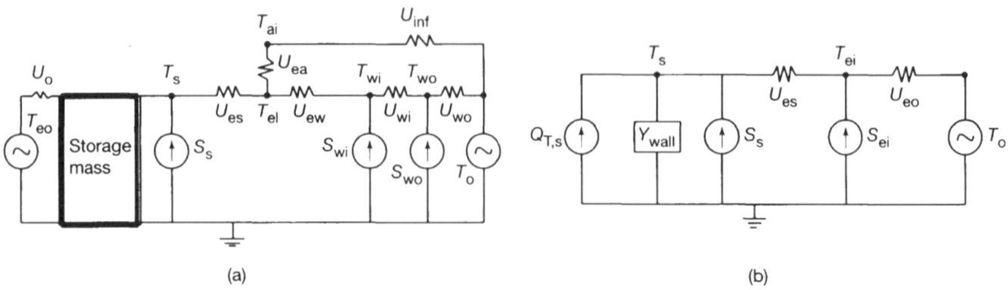

Figure 5.2 (a) Thermal network after delta-star transformation and (b) simplified network

Source S_{ei} represents the portion of solar radiation that is absorbed in the glazings and then transferred to the room interior by thermal radiation or convection. After obtaining S_{ei}, we can determine the equivalent conductance U_{eo} coupling T_{ei} and T_o as

$$U_{eo} = 1[(1/U_{ew} + 1/U_{wi} + 1/U_{wo}) + 1/(1/U_{ea} + 1/U_{inf})] \tag{5.4}$$

(If there is no mass on the ceiling, the ceiling is modelled as a pure conductance in parallel with U_{inf} and its conductance is added to U_{inf} in eqn 5.4.) Both S_{ei} and S_s are proportional to the solar irradiance of the window, so that S_{ei} is expected to be proportional to S_s. It was found that S_{ei} is typically about 4% of the magnitude of S_s (for standard double glazing and mid-latitudes in winter). As can be seen from Figure 5.2(b), S_{ei} can be transformed to node *s* and added to S_s. Thus, we multiply S_s by the appropriate transformation factor so as to include the effect of S_{ei}; this factor is equal to $[1+(S_{ei}/S_s)U_{es}/(U_{es} + U_{eo})]$ (an average value of the ratio S_{ei}/S_s can be determined from equation 5.3 in the middle of the design period).

The final simplification is replacement of the wall by its Norton equivalent, which is determined as explained in Section 4.2. Since we are only interested in the interior surface temperature T_s, unimportant nodes are eliminated by this transformation and the resulting network consists of only two-terminal components. These two simplifications yield the network in Figure 5.2(b).

As shown in Figure 5.2(b), the Norton equivalent consists of Y_{wall} (called the wall self-admittance) and the equivalent source $Q_{T,s}$. The Norton equivalent source is given by

$$Q_{T,s} = -Y_{seo}T_{eo} \tag{5.5}$$

where the transfer admittance Y_{seo} is given by

$$Y_{seo} = -A_s/[A_s\cosh(\gamma x)/U_o + \sinh(\gamma x)/(k\gamma)] \tag{5.6}$$

The wall self-admittance is given by

$$Y_{wall} = A_s[(U_o/A_s) + k\gamma E]/[(U_o/k\gamma A_s)E + 1] \tag{5.7}$$

where $E = \tanh(\gamma x)$.

Important design approximation: For rooms with well-insulated walls and a high amount of mass (more than 10 cm of concrete or equivalent) we can approximate the wall self-admittance with the following equation:

$$Y_{wall} \approx Ak\gamma\tanh(\gamma x) \tag{5.7a}$$

For very thick mass (more than 25 cm of concrete or equivalent) we may approximate the wall as a semi-infinite solid with admittance equal to:

$$Y_{wallsemi\text{-}inf} \approx Ak\gamma \tag{5.7b}$$

In both of the above cases the transfer admittance (for all frequencies apart from the mean) is negligible.

The solutions for T_s and T_{ei} are easily obtained from a heat balance at the two nodes (note that S_{ei}, a small source, is not explicitly included in the heat balance as it has been transformed to node s and added to S_s). The storage mass surface temperature is found to be given by

$$T_s(\omega) = (S_s + Q_{T,s} + U_L T_o)/Y_s \tag{5.8}$$

where Y_s is the room total admittance as seen from the port formed by the reference and the T_s node, given by

$$Y_s = Y_{wall} + U_L \tag{5.9}$$

where

$$U_L = (1/U_{es} + 1/U_{eo})^{-1}$$

The conductance U_L represents the total loss conductance between the storage mass interior surface and the ambient temperature T_o.

The representative temperature is found to be given by

$$T_{ei}(\omega) = (TF)_{eis}(S_s + Q_{T,s}) + (U_{eo}T_o)/Y_{ei} \tag{5.10}$$

where

$$(TF)_{eis} = 1/((Y_{wall}/U_L + 1)U_{eo}) \tag{5.11}$$

and

$$Y_{ei} = U_{eo} + Y_{wall}U_{es}/(Y_{wall} + U_{es}) \tag{5.12}$$

The transfer function $(TF)_{eis}$ determines the effect of the primary source S_s, while Y_{ei} is the total room admittance as seen from the port formed by the T_{ei} node and reference node.

5.2.3 Source models

Steady-state periodic conditions are assumed in computing the swing (difference between maximum and minimum) in T_{ei} or T_s over a particular day. The period is assumed to be one day, implying that the day in question has been preceded by identical days. Although deviations from the one-day periodic assumption (e.g. previous day overcast and current day clear) would cause errors in the *mean* value of T_{ei} for the day considered, the variation of the waveform of T_{ei} about its mean should not be affected and therefore the difference between its maximum and minimum points should be close to the actual swing. Another option would have been to include the effect of the previous day, i.e. to assume a two-day periodicity; the present method can be readily generalized to include such a model. Note that because of the one-day periodic assumption, weather input is needed only for the particular day in question.

Fourier series models are required for the sources S_s, T_{eo} and T_o, in order to obtain the time domain solution for T_s and T_{ei} under periodic conditions. For equatorial facing windows (e.g. south facing in the northern hemisphere) the primary source S_s can be closely modelled over the daytime as a half sinusoid. The constraints on the half sinusoid are chosen to satisfy two conditions: first that it has an integrated energy equal to the actual solar radiation H_a absorbed in the room over the day, and second that it must start at sunrise and terminate at sunset. Thus, S_s for a single day is defined by

$$S_s = 0, |t| > t_s \text{ and}$$
$$S_s = (\pi H_a/4t_s)\cos(\pi t/2t_s), |t| < t_s \tag{5.13}$$

where t is the time measured from the solar noon (t is negative in the morning, positive in the afternoon) and t_s is the value of t at sunset. This expression for S_s can be approximated by the following truncated Fourier series:

$$S_s = q_o + \sum_{n=1}^{N} q_n \cos(\omega_n t) \tag{5.14}$$

where $\omega_n = 2\pi n/t_d$, $q_o = H_a/t_d$ and $q_n = H_a f_n$, f_n being given by

$$f_n = \sin[\pi(t_d + 4t_s n)/(2t_d)]/(t_d + 4t_s n) + \sin[\pi(t_d - 4t_s n)/(2t_d)]/(t_d - 4t_s n) \tag{5.14a}$$

As can be seen, the *relative* magnitude of the harmonics is determined completely by the daylength $2t_s$. The number of harmonics necessary to model S_s was found to increase with decrease of the daylength because of the more abrupt increase of the absorbed radiation at sunrise, and its more rapid decrease at sunset. For windows that are not equatorial facing, the Fourier coefficients can be determined by direct numerical integration of the instantaneous absorbed irradiation. Various methods can be used for determining the total daily irradiation absorbed H_a. Either real weather may be employed, which is cumbersome, or instead the daily clearness index may be used to generate values of the hourly clearness index (Athienitis *et al.* 1986).

The variation of the ambient temperature T_o was modelled by a single sinusoid, with maximum at 3 p.m. and minimum at 3 a.m. (solar times). Thus, if T_{om} is the daily average of T_o, and ΔT_o the daily range of T_o then

$$T_o = T_{om} + (\Delta T_o/2)\cos(\omega_1 t - \pi/4) \tag{5.15}$$

The model does not distinguish between the differing orientations of the exterior wall surfaces and hence it is not possible to accurately model the effect of fluctuations in the solar radiation absorbed by the wall exterior surfaces. A sensitivity analysis given in a later section showed that the effect of this absorbed solar radiation on the temperature swing is small, provided there is at least a medium amount of insulation (about 5 RSI) behind the mass. Thus, the sol-air temperature T_{eo} can be modelled as equivalent to T_o.

5.2.4 Periodic solution

The time domain solution for the representative temperature T_{ei} is obtained after substituting the source models in equation 5.10 and evaluating magnitudes and phase angles for the complex transfer functions. Since we are only interested in fluctuations, the mean source terms can be ignored. The following solution is obtained for the variation of T_{ei} about its mean T_{eim}:

$$\begin{aligned} T_{ei}(t) - T_{eim} = {} & [\Delta T_o U_{eo}/(2|Y_{ei1}|)]\cos(\omega_1 t - \varphi_{ei1} - \pi/4) \\ & + (\Delta T_o |Y_{seo1}||(\mathrm{TF})_{eis1}|/2)\cos(\omega_1 t + \varphi_{eis1} + \varphi_{seo1} - \pi/4) \\ & + \sum_{n=1}^{N} |(\mathrm{TF})_{eisn}| q_n \cos(\omega_n t + \varphi_{eisn}) \end{aligned} \tag{5.16}$$

where $|(\mathrm{TF})_{eisn}|$ and φ_{eisn} are respectively the magnitude and phase of $(\mathrm{TF})_{eis}$ for $\omega = \omega_n$. Similar definitions apply for $|Y_{ei1}|$ and φ_{ei1}, and $|Y_{seo1}|$ and φ_{seo1}. For the variation of T_s we obtain:

$$T_s(t) - T_{sm} = [\Delta T_o U_L/(2|Y_{s1}|)]\cos(\omega_1 - \varphi_{s1} - \pi/4)$$
$$+ (\Delta T_{eo}|Y_{seo1}|/(2|Y_{s1}|))\cos(\omega_1 - \varphi_{s1} + \varphi_{seo1} - \pi/4)$$
$$+\sum_{n=1}^{N}(q_n / |Y_{sn}|)\cos(\omega_n t - \varphi_{sn}) \quad (5.17)$$

The swings in T_s and T_{ei} can be determined by differentiating the appropriate equation, finding the two zeros corresponding to the times at which the maximum and minimum temperatures occur, and then substituting these times back into the original equation.

Approximate design method for temperature swings: for a well-insulated room with massive walls, with equation 5.7a or 5.7b applicable, we can quickly estimate maximum temperature swings with the following approximation:

$$\text{Temperature swing} = 2F_h|q_1|/|Y_{wall}|$$

where q_1 is estimated from equation 5.14, $F_h = 1.12$ is a factor to approximate for the effect of higher harmonics, and $|Y_{wall}|$ may be estimated by

$$|Y_{wall}| = Ak|\gamma| \times |\tanh(\gamma x)|$$
$$= Ak(2\pi/(P\alpha))^{1/2}[(\cosh X - \cos X)/(\cosh X + \cos X)]^{1/2} \quad (5.17a)$$

with $X = 2L[\pi/(P\alpha)]^{1/2}$.

The phase angle of Y_{wall} is given by

$$\text{Phase}\ (Y_{wall}) = 45° + \text{atan}(\sin X/ \sinh X) \quad (5.17b)$$

Note that the first term (45°) in equation 5.17b corresponds to the phase angle for $k\gamma$ and is equivalent to a time lead of 3 h (15° is equivalent to 1 h for a period of 24 h).

5.2.5 Sensitivity analysis

The objectives of the sensitivity analysis are to study the relative significance of fundamental design variables and to document the effects of various modelling assumptions. It consists of two parts. The first analyses the transfer function's dependence on the resistance R_{ins}, the mass thickness x and the main heat loss conductance U_L. The second analyses the temperature swing's dependence on the above variables, plus the half-daylength t_s.

The assumed room data are: the dimensions of the south wall were 6.10 m wide by 2.44 m high, the north–south depth was 4.57 m, and the (south-facing, double-glazed) window area was 8.36 m^2. The infiltration value used was 0.5 air changes per hour. The heat transfer coefficients were as follows: $h_{cs} = h_{cw} = 3$ W/K/m^2,

h_i = 7.4 W/K/m^2 and h_{rw} = 5 W/K/m^2. The uniformly distributed storage mass was concrete with k = 1.73 W/K/m, ρ = 2242 kg/m^3 and c = 840 J/kg/K.

Three values were used for each of the four sensitivity variables: low (L), medium (M) and high (H); all are given in Table 5.1. The basis for choosing the thickness values shown relates to the non-dimensional thickness of the slab (Davies 1973) for the fundamental harmonic (X_1), which is equal to $(\omega_1/2\alpha)^{1/2}x$ (also shown in Table 5.1). For concrete, X_1 becomes equal to 6.3x and it is the most important quantity in determining Y_{wall} (equation 5.7) whose maximum magnitude is for approximately medium thickness.

Table 5.1 Variables for sensitivity analysis ($X_1 = (\omega_1/2\alpha)^{1/2}x$)

	x (m)	X_l	U_L *(W/K)*	R_{ins} *(m^2 K/W)*	t_s *(s)*
Low	0.08	0.50	27	1	14 000
Medium	0.14	0.88	37	5	20 000
High	0.20	1.26	47	10	26 000

SENSITIVITY OF TRANSFER FUNCTIONS

The most important transfer functions are Y_s and $(TF)_{eis}$ because they determine the effect of the primary source S_s on T_s and T_{ei} respectively. A sensitivity analysis is reported for Y_s but not $(TF)_{eis}$; because the latter is mainly a function of the same admittances as Y_s, similar results were found.

Room admittance Y_s. The room total admittance Y_s (equation 5.9) is the most important transfer function since it determines the effect of the absorbed solar radiation on the storage mass surface temperature. The variation of $|Y_{sn}|$ as a function of U_L, R_{ins} and x for six harmonics is shown in Table 5.2. Examination of Table 5.2 shows a low sensitivity to U_L: 27% increase in U_L causes a less than 1% increase in $|Y_{sn}|$. The effect of R_{ins} on $|Y_{sn}|$ is also small, the change being an increase

Table 5.2 Variation of room admittance magnitude $|Y_{sn}|$(W/K)

			Harmonic no. *n*					
R_{ins}	U_L	*x*	*1*	*2*	*3*	*4*	*5*	*6*
L	M	L	985	1863	2595	3176	3631	3990
		M	1494	2300	2735	3045	3314	3570
		H	1624	2154	2529	2878	3209	3513
	H	L	988	1867	2599	3181	3636	3997
		M	1498	2306	2742	3052	3322	3577
		H	1630	2162	2536	2886	3216	3524
M	M	L	1000	1895	2633	3213	3663	4017
		M	1530	2325	2746	3048	3314	3566
		H	1648	2158	2527	2877	3208	3517
H	M	L	1003	1900	2638	3218	3667	4020
		M	1536	2328	2748	3049	3314	3568
		H	1651	2158	2527	2877	3208	3517

of about 2% when R_{ins} changes from low to high. Not unexpectedly, the mass thickness x is the most important parameter. The magnitude $|Y_{sn}|$ increases with the harmonic number n due primarily to the increase of γ, corresponding to a decrease of the penetration depth $(\alpha/\omega)^{1/2}$.

Wall transfer admittance Y_{seo1}. The magnitude of the wall transfer admittance Y_{seo1} (first harmonic) determines the magnitude of the (single harmonic) transformed source $Q_{T,s}$ (equation 5.5) and thus the effect of the sol-air temperature. The variation of $|Y_{seo1}|$ is shown in Table 5.3 as a function of both R_{ins} and x. As expected, $|Y_{seo1}|$ is highest for low R_{ins} and low for x.

Both R_{ins} and x are important in determining $|Y_{seo1}|$. Comparing these values of $|Y_{seo1}|$ with the values of $|Y_{sn}|$ given in Table 5.2 shows that Y_{seo} is small compared to Y_s. Thus the effect of fluctuations in the sol-air temperature on swings of T_s or T_{ei} is expected to be small.

Room admittance Y_{ei}. A sensitivity analysis was also performed for Y_{ei} (equation (5.12)), whose magnitude $|Y_{ei1}|$ for the first harmonic determines the effect of the swing of the ambient temperature on T_e (see equation 5.16). As shown later, it is a relatively unimportant transfer function since the contribution which ΔT_o (range of T_o) makes to the variation in T_s is small. It was found that Y_{ei1} is more sensitive to changes in U_L than is Y_s; a change of U_L from low to high causes a maximum increase in $|Y_{ei1}|$ of approximately 5% (in changing U_L it was assumed that U_{eo} changes but not U_{es}; this is equivalent to assuming that the main causes of changes in U_L are changes in the infiltration loss and in h_i and h_o).

Table 5.3 Variation of $|Y_{seo1}|$(W/K)

x	R_{ins} *L*	R_{ins} *M*	R_{ins} *H*
L	83	18	9.0
M	71	15	7.8
H	51	11	5.6

SENSITIVITY OF MASS SURFACE TEMPERATURE SWINGS

The variation of the mass surface temperature T_s results from the highly important heat storage and release process. Since superposition applies, the temperature swings in T_s caused by each source may be studied separately.

Swings caused by absorbed solar radiation. An important objective of this analysis was to study how many harmonics are needed to model the effect of S_s. Using equation 5.17 the magnitude of each harmonic of T_s is given by $q_n|Y_{sn}|$. Thus, in order to study the importance of higher harmonics of S_s, a non-dimensional sensitivity function Sens(ω) was defined as the ratio of higher T_s harmonics to the fundamental:

$$\text{Sens}(\omega) = (q_n/|Y_{sn}|)/(q_1/|Y_{s1}|) \qquad (5.18)$$

The two variables essentially determining the sensitivity function are the daylength $2t_s$ and the storage mass thickness. Table 5.4 shows the results for low (L), medium (M) and high (H) value combinations of the two variables. As can be seen, the higher harmonics become more important for low mass and low daylength. This observation is further illustrated in Table 5.5, which shows how accurately the temperature swing would be determined using only a few harmonics.

For the other extreme case of high mass and high t_s, the fundamental harmonic would be enough for determining the temperature swings. The study indicates that between three to five harmonics should be sufficient to ensure accuracy of the swings of about 0.1 K.

Table 5.4 Sens(ω_n) (equation (5.18))

		Harmonic no. *n*				
x	t_s	*2*	*3*	*4*	*5*	*6*
L	L	0.385	0.150	0.036	–0.012	–0.019
	M	0.260	0.024	–0.027	–0.009	0.008
	H	0.126	–0.037	–0.006	0.011	–0.002
M	L	0.480	0.221	0.058	–0.020	–0.033
	M	0.324	0.035	–0.043	–0.015	0.014
	H	0.157	–0.054	–0.009	0.019	–0.004
H	L	0.557	0.258	0.066	–0.022	–0.036
	M	0.376	0.041	–0.049	–0.017	0.015
	H	0.182	–0.063	-0.011	0.021	–0.004

Table 5.5 Swings of T_s (K) due to S_s (H_a = 83.6 MJ)

		Number of harmonics used					
x	t_s	*1*	*2*	*3*	*4*	*5*	*6*
L	L	3.50	4.21	4.23	4.23	4.26	4.32
M	L	2.29	2.92	3.07	3.05	3.07	3.11
M	M	2.06	2.37	2.36	2.42	2.43	2.43
M	H	1.77	1.85	1.91	1.92	1.93	1.93
H	H	1.65	1.72	1.77	1.77	1.77	1.77

Swings due to ambient and sol-air temperatures. The ambient temperature T_0 causes swings of the room temperature by virtue of its variation ΔT_0. Its effect will be compared to that of the fundamental harmonic q_1 of the absorbed radiation, the latter accounting for about 70% of the swing of T_s caused by S_s. The ratio of the two effects is $U_L \Delta T_0/2q_1$. Assuming an extreme value for ΔT_0 equal to 20 K, a value for U_L of 40 W/K and for q_1 of 2000 W, the maximum value of this ratio is 0.2. This means that the maximum effect of T_0 on the total swing of T_s is at most 14% of the contribution of S_s.

The effect of T_{eo} on T_s can be studied by a similar order of magnitude analysis. The ratio of the effect of T_{eo} on T_s to the effect of T_o on T_s can be shown to be equal to the ratio $|Y_{seo1}|/U_L$, which, for the typical case of medium insulation and medium wall thickness, is less than 0.5. The difference between the maximum and minimum values of T_{eo} on a clear day is shown by Threlkeld (1970) to be a maximum of about 40 K for a horizontal surface (latitude = 42°N and surface absorptance = 0.9). Since only the roof is horizontal, the fluctuation of T_{eo} about its mean would be at most between 25 and 30 K, of which 20 K would be contributed by T_o thus, roughly speaking, on a clear day, the swings in T_{eo} will be roughly twice the swings in T_o. After correcting by the factor $|Y_{seo1}|/U_L$, we conclude that the contribution of T_{eo} to the swing in T_s will be about the same as that of T_o, or about 14% of the contribution of S_s. Since the externally absorbed solar radiation contributes only one half of the variation in T_{eo}, we conclude that the externally absorbed solar radiation provides a contribution to the swings in T_s equal to at most about 7% of the contribution of S_s. The contribution of the externally absorbed radiation can be shown to be even smaller if the effect of the phase shift introduced by Y_{seo} is considered. Thus, it is sufficient for this simplified model to assume T_{eo} equivalent to T_o and hence ignore the effect of the externally absorbed solar radiation.

5.3 Bibliography

Athienitis A.K., Sullivan H.F. and Hollands K.G.T. (1986) 'Analytical model, sensitivity analysis and algorithm for temperature swings in direct gain rooms', *Solar Energy*, **36**:4, 303–312.

Balcomb J.D. *et al.* (1980) *Passive Solar Handbook*, Vol. 2, Los Alamos Scientific Laboratory, USA.

Carslaw H.S. and Jaeger J.C. (1959) *Conduction of Heat in Solids*, 2nd Edn, Oxford University Press, London.

Davies M.G. (1973) 'The thermal admittance of layered walls', *Building Science*, **8**, 207–220.

Kirkpatrick A.T. and Winn C.B. (1984) 'Spectral analysis of the effective temperature in passive solar buildings', *Journal of Solar Energy Engineering*, **106**, 112–119.

Roe P.H. (1966) *Networks and Systems*, Addison Wesley Co.

Threlkeld J.L. (1970) *Thermal Environmental Engineering*, 2nd Edn, Prentice Hall, Englewood Cliffs, NJ.

Winn C.B. (1987) 'Control in solar energy systems', *Advances in Solar Energy*, ASES, pp. 209–240.

CHAPTER 6

Ventilation and indoor air quality

6.1 Airflow processes in buildings

Ventilation of buildings may be achieved by natural or mechanical means, in order to provide thermal comfort, humidity control and acceptable indoor air quality. In particular, ventilation is used to dilute indoor concentration of pollutants and renew building air. In parallel, thermal comfort depends to a certain degree on indoor air velocity and appropriate indoor air movement can provide thermal comfort, even when the temperature and humidity are not the most appropriate. If temperatures are high, too little air movement can cause discomfort, and higher air movement may be welcome. The sizing and the design of the ventilation system should consider the effect that the air movement in the room will have on comfort if the occupants are not to subvert the ventilation strategy. Existing standards for indoor air movement assume that high air velocities and turbulence may cause discomfort through draughts. Appropriate ventilation for humidity control may prevent microbiological growth on walls, floors and ceilings as well as protect the building construction from damage.

Mechanical ventilation can be used when a constant or controlled airflow is necessary or when outdoor conditions are not favourable to provide a healthy and comfortable indoor environment for building occupants. The provision of ventilation air to a mechanically ventilated building requires that outdoor air is 'cleaned' and conditioned before it enters the building, which may be an important cost in building operation. Cleaning of the circulating air through filters is well used in everyday practice; however, it removes only the coarse particles, which may not be so hazardous to health as fine particles, which are difficult to remove.

Recently, energy conservation measures have promoted the application of heat recovery from exhaust air and a higher recirculation of the return air. In some cases, they have led to the reduction of the quantities of outside air, which may have serious impact on indoor air quality. Recent ventilation standards consider both energy conservation and indoor air quality and propose methodologies to optimize the ventilation strategy; however, the level of requirements between the various proposed standards varies considerably. In parallel there is an uncertainty regarding the IAQ requirements. For example in the CEN approach, the uncertainty in the real indoor air quality needs varies by 250%: from 4 to 10 l/s per person in relation to occupants' related pollution. If one takes into consideration smoking, the variation is even larger (see Table 6.1).

Table 6.1 Ventilation requirements for the various building categories

Category	Required ventilation rate (l/s per occupant) *No smoking*	*20%*	*40%*	*100%*
A	10	20	30	30
B	7	14	21	21
C	4	8	12	12

It is certain that in practice the ventilation performances do not satisfy the specifications found in the standards and regulations. To achieve ventilation systems with good performances with respect to indoor air quality and energy efficiency, it is important that the system is designed appropriately, while the performance of the different components as well as the quality of execution is of a correct level. Therefore, it is important to consider appropriate mechanisms for achieving a better compliance.

When the local climate and the concentration of the outdoor pollutants are not restrictive factors, the use of natural ventilation techniques may be an efficient and energy-saving procedure to maintain an appropriate indoor air quality, achieve thermal comfort and provide free cooling. Driving forces of natural ventilation are wind and the thermal difference between indoor and outdoor air. Possible problems related to the use of natural ventilation techniques involve the lack of particle filtration; possible poor indoor temperature control; exposure to outdoor noise; and probably low ventilation rates under adverse weather conditions.

6.2 Natural ventilation

6.2.1 Airflow in urban areas

Airflow around isolated buildings is well known and is characterized by a bolster eddy vortex due to flow down the windward façade. Behind there is a lee eddy drawn into the cavity of low pressure due to flow separation from the sharp edges of the building top and sides, and further downstream is the building wake characterized by increased turbulence but lower horizontal speeds than the undisturbed flow.

The airflow patterns in urban canyons have received important attention over the past years (Table 6.2). Natural ventilation of buildings located in urban canyons is seriously reduced because of important decrease of the wind velocity inside the canyons.

Table 6.2 Airflow in urban canyons

Flow along the canyon	*Perpendicular flow*	*Oblique flow*
Wind speed above the canyon < 4 m/s		
Type of flow: chaotic. There is no coupling between the undisturbed and the wind speed inside the canyon. The maximum expected wind speed should not exceed 0.5 m/s. This value may be used in ventilation models especially when the ambient speed exceeds 3 m/s and the H/W ratio is low (<1). For lower speeds a value close to 0.2–0.3 m/s may be used. There is no predominant wind direction inside the canyon. When the ambient wind speed exceeds 3 m/s, it may be considered that the flow is almost parallel to the axis of the canyon with a very small uplift component parallel to the walls of the canyon.	**For $H/W > 0.7$** **Type of flow: Chaotic**. There is no coupling between the undisturbed and the wind speed inside the canyon. Thermal as well as mechanical influences play an important role in the canyon circulation. The maximum expected wind speed should not exceed 0.4 m/s. This value (0.4 m/s) may be used in ventilation models especially when the ambient speed exceeds 3 m/s. For lower speeds a value close to 0.2 m/s may be used. There is no predominant wind direction inside the canyon.	Similar to the perpendicular flow.
Wind speed above the canyon > 4 m/s		
Type of flow: Mean wind along the canyon axis with possible uplift along the canyon walls. **Prediction of the along-canyon wind speed $v(z)$** Use: $v(z) = U_o e^{z/Z_o}$ where z is the height, U_o is a constant reference speed (see Figure 6.1) and Z_o is the roughness length of the obstructed sublayer calculated by the expression: $Z_o = h_b D^*/z_o$ where D^* is an effective diameter of air space between obstacles and can	**For canyons with $H/W < 2$ and $L/W > 20$** **Type of flow:** A stable circulatory vortex is established in the canyon. **Prediction of the cross-canyon u and vertical, w, air speeds:** $u = u_0(1 - \beta)^{-1}$ $[\gamma(1 + k\gamma) - \beta(1 - k\gamma)/\gamma]\sin(kx)$ and $w = -u_0(1 - \beta)^{-1} k y[\gamma - \beta/\gamma] \cos(kx)$ where $k = \pi/W$, $\beta = \exp(-2kH)$,	**For canyons with $L/W > 20$** **Type of flow:** A spiral vortex is induced along the length of the canyon, a corkscrew type of action. **Prediction of the cross-canyon u and vertical, w, air speeds:** $u = u_o(1 - \beta)^{-1} [\gamma(1 + k\gamma)$ $- \beta(1 - k\gamma)/\gamma]\sin(kx)$ and $w = -u_0(1 - \beta)^{-1} k y[\gamma - \beta/\gamma]\cos(kx)$ where $k = \pi/W$,

Table 6.2 (continued)

<table>
<tr>
<td>be tentatively approximated for the city by:

$D^* = 0.1h_b$

The parameter h_b is equal to the height of buildings.
Typical values of z_o are given by Oke (1987); see Table 6.3.
Alternatively $v(z)$ may be calculated by the following expression proposed by Yamartino and Wiegard (1986):

$v(z) = v_r \log[(z + z_o)/z_o]/\log[(z_r + z_o)/z_o]$

where v_r is the value at reference height z_r and z_o is the surface roughness.
For canyons with L/W < 20
Type of flow: Intermittent vortices are shed on the building corners. These vortices are responsible for the mechanism of advection from the building corners to mid-block creating a convergence zone in the mid-block region of the canyon.
Prediction of the wind speed: Use mean wind speed close to 0.4 m/s. Prevailing direction: Parallel to the axis of the axis with a downward direction.</td>
<td>$\gamma = \exp(ky)$, $y = z - H$, and u_0 is the wind speed above the canyon and at the point $x = W/2, z = H$.
For canyons with H/W > 2 and L/W > 20
Type of flow: Two vortices are developed, an upper one driven by ambient airflow and a lower one driven in the opposite direction by the circulation above.
Prediction of the cross-canyon u and vertical, w, air speeds:
For the lower part of the canyon use a mean wind speed close to 0.3 m/s.
For the upper parts use the following expressions:

$u = u_o(1 - \beta)^{-1}[\gamma(1 + k\gamma) - \beta(1 - k\gamma)/\gamma]\sin(kx)$

and

$w = -u_o(1 - \beta)^{-1}ky[\gamma - \beta/\gamma]\cos(kx)$

where $k = \pi/W$, $\beta = \exp(-2kH)$, $\gamma = \exp(ky)$, $y = z - H$ and u_o is the wind speed above the canyon and at the point $x = W/2, z = H$.
For canyons with L/W < 20
Type of flow: Intermittent vortices are shed on the building corners. These vortices are responsible for the mechanism of advection from the building corners to mid-block creating a convergence zone in the mid-block region of the canyon.
Prediction of the wind speed: Use mean wind speed close to 0.4 m/s. Prevailing direction: Parallel to the axis of the axis with a downward direction.</td>
<td>$\beta = \exp(-2kH)$, $\gamma = \exp(ky)$, $= z - H$ and u_o is the wind speed above the canyon and at the point $x = W/2, z = H$.
For canyons with L/W < 20
Type of flow: Intermittent vortices are shed on the building corners. These vortices are responsible for the mechanism of advection from the building corners to mid-block creating a convergence zone in the mid-block region of the canyon.
Prediction of the wind speed: Use mean wind speed close to 0.4 m/s. Prevailing direction: Parallel to the axis of the axis with a downward direction.</td>
</tr>
</table>

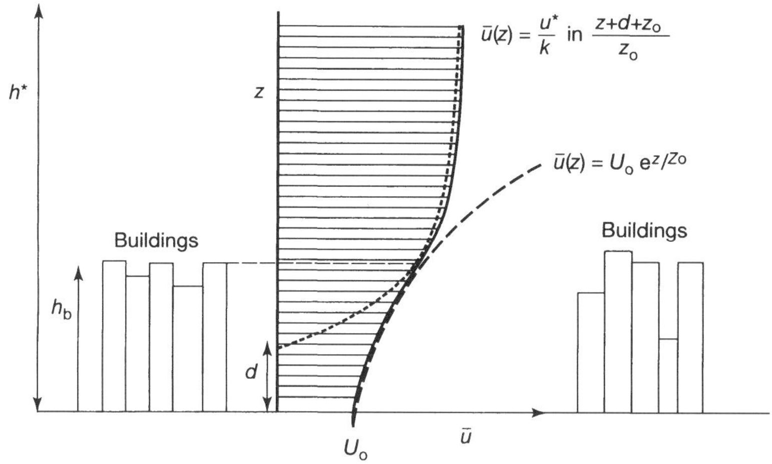

Figure 6.1 Logarithmic and exponential wind profiles in the surface layer, above and below building height (after Nicholson 1975)

Table 6.3 Typical roughness length z_0 of urbanized terrain

Terrain	z_0 *(m)*
Scattered settlement	0.2–0.6
Suburban	
Low-density residences and gardens	0.4–1.2
High-density residences and gardens	0.8–1.8
Urban	
High density, more than five-storey row and block buildings	1.5–2.5
High density plus multistorey blocks	2.5–10

From Oke (1987)

Natural ventilation of buildings is due either to the wind forces or to the temperature difference between the indoor and outdoor environment or to a combination of both. Design of urban buildings to improve natural ventilation potential should consider the appropriate wind data and not routine meteorological observations collected in open fields. Also, the specific temperature regime in a canyon should be considered.

Knowledge of the air speed inside urban canyons is of great importance for passive cooling applications and especially for naturally ventilated buildings. Various methods, simplified or detailed, have been proposed to calculate the wind speed inside a canyon. A detailed presentation of the existing methods to calculate the wind speed in urban canyons has been presented in Santamouris (2000).

Urban canyons are characterized by three main parameters: *H*, the mean height of the buildings in the canyon, *W*, the canyon width and *L*, the canyon length. Given these parameters, the geometrical descriptors are limited to three simple measures.

These are the ratio H/W, the aspect ratio L/H and the building density $j = A_r/A_l$ where A_r is the plan of roof area of the average building and A_l is the 'lot' area or unit ground area occupied by each building.

When the predominant direction of the airflow is approximately normal (say 30°) to the long axis of the street canyon, three types of airflow regime are observed as a function of the building (L/H), and canyon (H/W) geometry. When the buildings are well apart ($H/W > 0.05$), their flow fields do not interact. At closer spacing, the wakes are disturbed and the flow regime is known as **isolated roughness flow**. When the height and spacing of the array combine to disturb the bolster and cavity eddies, the regime changes to one referred to as **wake interference flow**. This is characterized by secondary flows in the canyon space where the downward flow of the cavity eddy is reinforced by deflection down the windward face of the next building downstream. At even greater H/W and density, a stable circulatory vortex is established in the canyon because of the transfer of momentum across a shear layer of roof height, and transition to a 'skimming' flow regime occurs where the bulk of the flow does not enter the canyon.

Skimming regime is the most common in urban areas. Under these conditions the airflow in the canyon can been seen as a secondary circulation feature driven by the above roof-imposed flow. If the wind speed out of the canyon is below some threshold value the coupling between the upper and secondary flow is lost, and the relation between wind speeds above the roof and within the canyon is characterized by a considerable scatter. According to many studies carried out in almost symmetrical canyons where $1 < H/W < 1.4$, the threshold value is between 1.5 and 2 m/s. In all these studies higher wind speeds have been found to produce a stable vortex circulation within the canyon. For lower wind speeds thermal as well as mechanical influences may play an important role in the canyon circulation.

Parallel ambient flow generates a mean wind along the canyon axis, with possible uplift along the canyon walls as airflow is retarded by friction by the building walls and street surface. Regarding the relation between the free stream wind speed, U, and the along-canyon velocity, v, it is reported that the along-canyon wind component, v, in the canyon is directly proportional to the above-roof along-canyon component, through the constant of proportionality that is a function of approach flow azimuth.

The more common case in the urban environment is that where airflows at a certain angle are relative to the long axis of the canyon. Unfortunately, the existing research on this topic is considerably less than the scientific information for perpendicular and along-canyon flows. However, it is known that when the flow above the roof is at some angle of attack to the canyon axis, a spiral vortex is induced along the length of the canyon, similar to a corkscrew type of action. For intermediate angles of incidence to the canyon long axis, the canyon airflow is the product of both the transverse and parallel components of the ambient wind, where the former drives the canyon vortex and the latter determines the along-canyon stretching of the vortex.

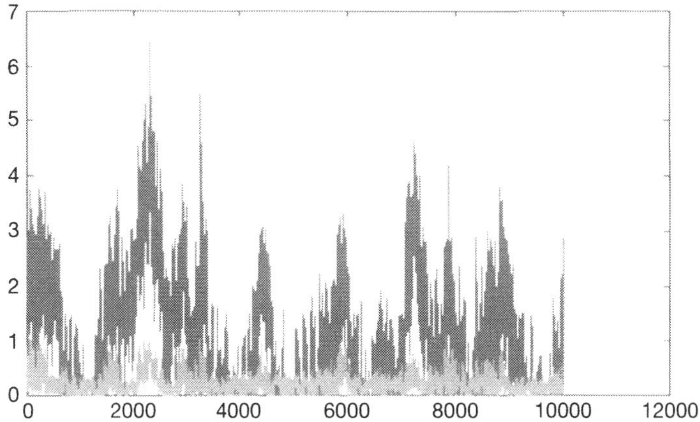

Figure 6.2 Wind velocity inside the canyon (grey) and above the canyon (black)

As a result of the above mechanisms, wind speed inside the canyons in an urban environment is seriously decreased and as a mean value is close to 10–30% of the undisturbed wind speed over the buildings. Figure 6.2 shows the values of the wind speed measured over and inside the canyon in an urban canyon in Athens. As shown, although ambient wind speeds easily exceed 4 m/s, the speed inside the canyon increases up to 1 m/s, and rarely exceeds it.

Decrease of wind speed seriously affects the potential for natural ventilation of buildings in the urban environment. For better design of natural ventilation in urban buildings, different methods have been proposed to consider realistic wind speeds. In fact, various methods, simplified and detailed, have been proposed to calculate the effects of the building location on the natural ventilation effectiveness. The methods in general propose techniques to reduce either the wind speed or the airflow rate due to the relative location of the building and the characteristics of the neighbourhood buildings.

A detailed method to calculate or select the appropriate wind speed in the urban environment is presented below. The methodology classifies the undisturbed wind flow in three main classes: flow (a) parallel to the canyon axis, (b) perpendicular and (c) oblique to the axis of the road. It classifies canyons as a function of their height (H) to width (W) ratio, as well as of the ratio of the canyon length (L) to the canyon width (W). The methodology considers that when the ambient wind speed above the canyon is lower than 3–4 m/s, there is no clear coupling between the flow above and inside the canyon, while it takes into account advection phenomena and end effects from the canyon corners. For these cases empirical wind speeds are proposed.

For higher wind speeds, and when the coupling between the ambient and the canyon wind speeds is established, the model proposes using the Hotchkiss and Harlow (1973) methodology to calculate the canyon transverse air speed $(w^2 + u^2)^{0.5}$, for perpendicular or oblique to the canyon axis winds, where u and w are the cross-canyon and vertical air speeds respectively. This model considers incompressible flow, absence of sources or sinks of vorticity within the canyon, and appropriate boundary conditions for the simple $2 - d$ rectangular notch of depth H and width W. For wind speeds parallel to the canyon axis and when coupling between the ambient and canyon wind speeds is established, the proposed methodology is based on the algorithms proposed by Yamartino and Wiegnard (1986).

6.2.2 Natural ventilation calculation methodologies

Transfer of air between the building and the surrounding environment as well as among the various zones of the buildings plays a very important role in the overall thermal balance of buildings, which also affects the levels of thermal comfort and indoor air quality. Exchange of air can be achieved either by mechanical means (mechanical ventilation) or through large openings of the building's envelope (natural ventilation). Air is also exchanged in an uncontrolled manner through cracks and other small size openings (infiltration).

Knowledge of the exact air supply to a building is necessary to determine its thermal performance and the concentration of the indoor pollutants. Classical fluid dynamics provides the theoretical framework to calculate the airflow when mechanical ventilation systems are used. This can be achieved by using either computerized fluid dynamic techniques (CFDs), or more simplified network models. Calculation of airflow through large openings, natural ventilation, is a more complicated task. The random nature of the wind makes estimation of airflow characteristics much more complicated than in the case of mechanical ventilation. Computerized flow dynamic and network tools have been proposed to treat natural ventilation problems. These tools are more efficient when airflow is due to temperature differences and not to wind. However, the use of complicated CFD techniques is not the more suitable solution, especially for small and medium projects. The basic principles as well as details of simplified, network and CFD calculation models are discussed in Allard (1998).

Computerized fluid dynamics tools are based on the solution of the Navier–Stokes equations combined with turbulence models. These tools determine the field of air velocity in a zone, the flow rate through various components, and the concentration of indoor pollutants in zones. Also, they combine efficient heat transfer models to calculate the thermal balance of various building elements. Although these tools are very powerful, only experts are able to use them while their cost is high.

Network calculation models are based just on the equation of mass conservation combined with some empirical knowledge. These tools can efficiently simulate the airflow rate in a building, when natural ventilation techniques are used, as well as the concentration of indoor pollutants, but they do not provide any information on the velocity field in a space. Network models are easy in their use, and their cost is minimal. Some of the models, such as the AIOLOS software (Allard 1998), are free to the public.

In network models, the flow rate (Q) through an opening of relatively large free area is calculated using the common orifice flow equation:

$$Q = C_d A\sqrt{(2\Delta P/\rho)} \tag{6.1}$$

where

C_d = the discharge coefficient of the opening

A = the opening area in m^2

ΔP = the pressure difference across the opening in Pa

ρ = the air density in kg/m^3

For the calculation of the total pressure difference across the opening, the terms of the dynamic pressure due to the wind must be added to those representing the stack effect. Thus:

$$\Delta P = P_{1,0} - P_{2,0} + \rho_1 C_p V_1^2/2 - \rho_2 C_p V_2^2/2 + (\rho_1 - \rho_2)gz \tag{6.2}$$

where V_1, V_2 are air velocities at the two sides of the opening and at height z, C_p = pressure coefficient (wind) and g = acceleration due to gravity.

The zonal method is based on the concept that each zone of a building can be represented by a pressure node. Boundary nodes are also used to represent the environment outside the building. Nodes are interconnected by flow paths, such as cracks, windows or doors, to form a network.

Pressure at boundary nodes is known but has to be determined for internal nodes. Application of mass balance equation on a zone i with j flow paths gives:

$$\sum_{i=1}^{j} \rho Q_i = 0 \tag{6.3}$$

A multi-zone problem involving N zones is therefore described by a set of N equations with an equal number of unknown pressures. The application of mass balance on each internal node of the network leads to a set of simultaneous non-linear equations, the solution of which gives the internal node pressures. The flow rates through the building openings can then be determined as discussed previously.

Empirical manual methods have been also proposed to calculate airflow through large openings (natural ventilation). Some of them are the British Standard method (1980) and the manual algorithms of NORMA (Santamouris 1996). These tools provide an estimate of the airflow but should always be used within the limits of their validity.

6.2.3 Sizing of windows for natural ventilation purposes

Sizing of windows in a building is a compromise between various parameters defining mainly the thermal, visual and energy performance of the building. To determine the optimum size of the openings in a building to achieve a specific ventilation performance, any of the above methodologies, and in particular any network model, may be used.

Various simplified methods to size windows have been proposed. A summary of the main characteristics and limitations of six simplified empirical opening sizing methodologies is presented in Table 6.4.

Table 6.4 Characteristics of six simplified empirical methodologies to design openings for natural cross ventilation of buildings

Method	*Openings characteristics*	*Flow forces*
The Florida Solar Energy Center Method – I	Considers equal inlets and outlets. Calculates the gross opening area. Considers a screen with a porosity factor of 0.6.	Considers only the wind forces. Proposes correction factors for the orientation of the wind, terrain type, neighbouring buildings and height of openings.
The Florida Solar Energy Center Method – II	Considers non-equal inlets and outlets. It calculates an effective window area. Proposes a methodology to account for screens and window porosity.	Considers only the wind forces. Takes into account the pressure coefficients due to wind. Proposes correction factors for the neighbouring buildings and the height of openings.
The ASHRAE Method	Considers non-equal inlets and outlets. Proposes a coefficient to account the effectiveness of the opening as a function of the incidence angle of the wind.	Considers wind or temperature difference effects. Does not consider combined effects of wind and temperature difference.
Simplified Method of the University of Athens	Considers non-equal inlets and outlets. Valid for ratio inlets/outlets lower than 2.	Considers only the wind forces. Takes into account the pressure coefficients due to wind.
Aynsley Method	Considers non-equal inlets and outlets.	Considers only the wind forces. Takes into account the pressure coefficients due to wind.
British Standards Institution Method	Considers non-equal inlets and outlets.	Considers both the effect of the wind and temperature difference. It proposes criteria for combined effects.

A simplified but efficient methodology for windows sizing is proposed by the Florida Solar Energy Center (Chandra *et al.* 1986). The method assumes that the inlet and outlet areas are equal. The method can be used for slight differences; e.g. inlet = 40% of the total area. It does not account for airflows due to temperature difference. For a two-storey building the calculations should be done for each floor.

According to the method, the required gross total opening area, TOA, inlets plus outlets, in order to achieve a certain number of air changes per hour, ach, can be calculated by the following expression:

$$\text{TOA} = 0.00079V(\text{ach})(Wf_1 f_2 f_3 f_4) \tag{6.4}$$

where
TOA = total opening area in ft^2
V = volume of the building in ft^3
ach = design air change rate per hour
W = wind speed (mph), as measured by the nearest meteorological station
f_1 is the inlet to site 10 m windspeed ratio. This coefficient is a function of the wind incidence angle and is given in Table 6.5. Data are mainly obtained from Vickery and Baddour (1983), wind tunnel data. Incidence angle is zero when winds are perpendicular to a building face.

Table 6.5 f_1 inlet to site 10 m windspeed ratios

Wind incidence angle (°)	f_1
0–40	0.35
50	0.30
60	0.25
70	0.20
80	0.14
90	0.08

f_2 is the terrain correction factor. It is a function of the building location and the ventilation strategy. It can be obtained from Table 6.6.

Table 6.6 f_2: terrain correction factor

Terrain type	f_2: *24 h ventilation*	f_2: *night only ventilation*
Oceanfront or >3 miles (2 km) water in front	1.30	0.98
Airports, or flatlands with isolated-wall separated buildings	1.00	0.75
Rural	0.85	0.64
Suburban or industrial	0.67	0.50
Centre of large city	0.47	0.35

f_3 is the neighbourhood correction factor. This coefficient is a function of the wall height of the upwind building, h, as well as of the gap between the building and the adjacent upwind building, g. Values of f_3 are given in Table 6.7 as a function of the ratio g/h, and are extrapolated data from wind tunnel results by Lee *et al.* (1980).

Table 6.7 f_4: neighbourhood correction factor

Ratio g/h	f_3
0	0.00
1	0.41
2	0.63
3	0.77
4	0.85
5	0.93
6	1.00

f_4 is a height multiplication factor. If sizing windows for the second floor or for house on stilts, f_4 is equal to 1.15. Otherwise, use f_4 =1.

Example

For a given location W = 8.8 mph. Also, V = 10672 ft^3, ach = 30, while the incidence angle is 10°. The building is located in a suburban area and is ventilated on a 24 h basis. The wall height of the upwind building, h = 8 ft, while g = 24 ft. Use f_4 = 1. Determine the total opening area.

From Tables 6.5–6.7 we get: f_1 = 0.35, f_2 = 0.67, f_3 = 0.77. Then from the previous equation we get:

$$\text{TOA} = 0.00079V(\text{ach})/(\text{W}f_1 f_2 f_3 f_4)$$

$$\text{TOA} = (0.00079 \times 10\ 672 \times 30)/(8.8 \times 0.35 \times 0.67 \times 0.77 \times 1)$$

$$\text{TOA} = 159.1\ \text{ft}^2$$

For widely different inlet and outlet areas, use of the method developed by the same authors (Chandra *et al.* 1983) is proposed. This method is a simple procedure for sizing inlet and outlet window areas in cross-ventilated rooms. The method is based on the pressure difference coefficient across the inlet and outlet and permits the calculation of an effective window area, A, which is a combination of the open inlet and outlet areas. The method does not account for the airflow due to temperature difference between indoors and outdoors. The proposed procedure is for rooms with one effective inlet and one effective outlet; all the inlet windows are assumed to experience identical positive pressures and all outlet windows to experience identical negative pressures.

A is defined by the following expression:

$$A = A_o A_i / (A_o + A_i^2)^{0.5} \qquad (6.5)$$

where A_o and A_i are the open outlet and inlet respectively in ft^2.

To calculate A, the method proposes the following expression:

$$A = 0.000296V(\text{ach})/[Wf_1(f_3 f_4 \text{PD})^{0.5}]$$

where the parameter PD is the pressure difference coefficient acting across the inlet and outlet given by:

$$\text{PD} = \text{WPC} - \text{LPC} \qquad (6.6)$$

where WPC is the windward pressure coefficient and LPC the leeward pressure coefficient. The method proposes to use pressure coefficients given by Vickery and Baddour (1983) and Cermak *et al.* (1981) for the four facades of a residential building (see Table 6.8). Recommended pressure coefficient, PC, values for other apertures are:

- inlet with wingwall assist, PC = 0.40
- outlet with wingwall assist, PC = –0.25
- roof outlets, e.g. Venturi type, PC = –0.30.

Table 6.8 Pressure coefficients as a function of the wind incidence angle

Wind incidence angle, φ (°)	*PC at surface a*	*PC at surface b*	*PC at surface c*	*PC at surface d*
0.0	0.40	–0.40	–0.25	–0.40
22.5	0.40	–0.06	–0.40	–0.60
45.0	0.25	0.25	–0.45	–0.45
67.5	–0.06	0.30	–0.55	–0.40
90.0	–0.4	0.40	–0.40	–0.25

Source: Chandra *et al.* (1983)

To account the effect of blockage due to insect screens, partially open windows, etc., a porosity factor, PF, defined as the product of an insect screening porosity factor, IPF, with a window porosity factor, WPF, is proposed:

$$\text{PF} = \text{IPF} \times \text{WPF}$$

Values of IPF and WPF for various type of screens and windows are given in Tables 6.9 and 6.10. Then, the total, not open, inlet and outlet areas, TA_i, TA_o, are given by the expressions:

$$TA_i = A_i/\text{PF}$$

$$TA_o = A_o/\text{PF}$$

Table 6.9 Insect screening porosity factors, IPF

Screen type	*Typical IPF*
No screen	1.00
Bronze, 14 wires/inch	0.80
Fibreglass, 18 wires/inch	0.60

Source: Chandra *et al.* (1983)

Table 6.10 Window porosity factors, WPF

Window type	*Typical WPF*
Single or double hung	0.40
Awning, hopper, jalousi or projection that swivels open on a horizontal joint	0.60
Casement depending on fixed sash amount. Need to measure porosity	Varies

Source: Chandra *et al.* (1983)

Example

The volume of a building is 1536 ft^3, the required air changes per hour are 30, the wind speed is 8.8 mph, and the incidence angle on windward face, a, is 45°, while there is a wing wall assisted outlet at surface d. The building is in an urban area. The ratio g/h is equal to 3 (see previous example). Calculate: (a) the required open effective area A, (b) the inlet and outlet surface, if equal inlet and outlets are desired, and (c) the total, not open, inlet and outlet window areas if fibreglass screen and awning windows are considered.

(a) From Tables 6.5 and 6.6 we get: $f_1 = 0.67$, $f_3 = 0.59$, $f_4 = 1$. Also, from Tables 6.8–6.10 WPC = 0.25 and LPC = –0.25, then PD = 0.5. Then, using the previously given expressions, we get:

$$A = 0.000296V(\text{ach})/[Wf_1(f_3f_4\text{PD})^{0.5}]$$

$$= (0.000296 \times 1536 \times 30)/[(8.8 \times 0.67 \times 0.59^{0.5} \times 0.5)^{0.5}]$$

$$= 4.26 \text{ ft}^2$$

(b) If $A_o = A_i$ then $A_i = 1.41A$. Therefore, $A_i = A_o = 6.00$ ft^2.

(c) From Tables 6.8–6.10 we get: IPF = 0.8 and WPF = 0.60, then PF = 0.48. Therefore, $TA_i = TA_o = 6.00/0.48 = 12.5\ ft^2$.

6.2.4 Ventilation for indoor air quality

Indoor environmental quality can be seen as a combination of acceptable indoor air quality together with satisfactory thermal, visual and acoustic comfort conditions. Outdoor pollution is one of the sources of the so-called 'sick building syndrome'; others are related to indoor sources. Numerous studies have shown the serious impact of the outdoor environment on indoor air quality (Godish 1989). Solutions to indoor air pollution problems are source control, avoiding or attenuating the emission of contaminants, air cleaning and appropriate use of ventilation.

Inadequate ventilation may be the primary cause of poor indoor air quality in buildings. Monitoring of 356 public access buildings has shown that in approximately 50% of the buildings improper ventilation rates were the primary cause of illness complaints and poor air quality (Wallingford and Carpenter 1986). Ventilation problems have been exacerbated during the past years by certain energy-conserving measures. The reduction or elimination of fresh air, the reduction of infiltration and exfiltration, the lowering of thermostats in winter and the raising in summer, the elimination of the humidification or dehumidification systems and the early shut-down and late start-up of ventilation systems are some of the measures contributing to poor indoor air quality (Godish 1989).

The main objective of any ventilation system should be to provide good indoor air quality and satisfaction for the occupants (European Common Research Centre 1996). In addition, indoor air quality should protect the health of the occupants, promote productivity and protect the building, installations and furnishings.

As already mentioned, the more appropriate ventilation strategies are those that save energy and improve indoor air quality in parallel. Seppänen (1999) has summarized the more appropriate ventilation technologies and strategies for indoor air quality (see Table 6.11).

Inappropriate ventilation and indoor air quality may have a serious impact on the occupants' health, human well-being, employee performance and productivity at work. Improving indoor air quality in buildings is probably one of the most profitable investments. Seppänen and Palonen (1998) have estimated the total effects of poor indoor climate on national economy in Finland to be approximately 18 billion FIM.

Table 6.11 Summary of ventilation strategies for better indoor air quality and energy efficiency

Strategy or technology	*Phase of the technology*	*Indoor air quality*	*Estimated energy savings in ventilation*
Target values for indoor climate	Available	Improved	Up to 60% depending on climate
Particulate filtration of intake air	Available	Improved	No effect
Chemical air cleaning	Developing	Improved	Up to 100%
Balancing of airflows	Available	Improved	10–20%
Better ventilation efficiency	Available/ developing	Improved	Up to 50 %
Location of air intakes	Available	Improved	No effect
Heat recovery for			
large buildings	Available	No effect	Up to 70%
houses	Available	No effect	Up to 50%
Demand controlled ventilation	Available/ developing	No effect	Up to 30–40%, up to 50% in large spaces
Control of specific pollution sources	Available	Improved	Depends on application
Control of material emissions	Developing	Improved	Up to 50%
Task ventilation	Available/ developing	Improved	10–30%
Local exhausts	Available	Improved	Depends on application
Natural ventilation and free cooling	Available/ developing	Improved	Up to 60%
Operation and maintenance	Available/ developing	Improved	High (depends on initial level)

6.3 Bibliography

Allard F. (ed.) (1998) *Natural Ventilation in Buildings*, James and James (Science Publishers) Ltd, London.

BSI (1980) BS5925, 'Code of practice for design of buildings: Ventilation principles and designing for natural ventilation', British Standards Institution, London.

Cermak J.E. *et al.* (1981) 'Passive and hybrid cooling developments: natural ventilation – a wind tunnel study', Colorado State University, Fluid Mechanics and Wind Engineering Program Report No. CER81-82JEC-JAP-55A-MP24.

Chandra S., Fairey P.W. and Houston M.M. (1983) 'A handbook for designing ventilated buildings', Florida Solar Energy Center, Final Report, FSEC-CR-93-83, Cape Canaveral, FL.

Chandra S., Fairey P.W. and Houston M.M. (1986) 'Cooling with ventilation', SERI/SP-273-2966, DE86010701, Solar Energy Research Institute, Golden, CO.

European Common Research Centre (1996) 'European concerted action on indoor air quality and its impact on man', Report on ventilation and indoor air quality, European Common Research Centre, Ispra, Italy.

Godish T. (1989) *Indoor Air Pollution Control*, Lewis Publishers.

Hotchkiss R.S. and Harlow F.H. (1973) *Air Pollution Transport in Street Canyons*, Report by Los Alamos Scientific Laboratory for US Environmental Protection Agency, EPA-R4-73-029, NTIS PB-233 252.

Lee B.E., Hussain M. and Soliman B. (1980) 'Predicting natural ventilation forces upon low-rise buildings', *ASHRAE Journal*, February.

Nicholson S.E. (1975) 'A pollution model for street-level air', *Atmospheric Environments*, **9**, 19–31.

Oke T.R. (1987) *Boundary Layer Climates*, Cambridge University Press, Cambridge.

Santamouris M. (ed.) (2000) *Energy and Climate in the Urban Built Environment*, James and James (Science Publishers), London.

Santamouris M. (1996) 'NORMA – A method to calculate the thermal performance of passively cooled buildings', Vol. 5: Cooling Load of Buildings. Version 1.1.

Seppänen, O. (1999) 'Ventilation strategies for good indoor air quality and energy efficiency', *Proc. IAQ & Energy Conference*, Louisiana, ASHRAE, 1998.

Seppänen O. and Palonen J. (1998) 'The effects of indoor climate on national economy in Finland in billion FIM', FISIAQ, SIY Rapportti 10.

Vickery B.J. and Baddour R.E. (1983) 'A study of the external pressure distributions and induced internal ventilation flows in low rise industrial and domestic structures', University of Western Ontario, Boundary Layer Wind Tunnel Laboratory Report No BLWT-SS2-1983.

Wallingford K.M. and Carpenter J. (1986) 'Field experience overview: investigating sources of indoor air quality problems in office buildings', *Proc. IAQ 86*, ASHRAE, Atlanta.

Yamartino R.J. and Wiegnard G. (1986) 'Development and evaluation of simple models for the flow, turbulence and pollution concentration fields within an urban street canyon', *Atmospheric Environment*, **20**, 2137–2156.

CHAPTER 7

Sizing of small auxiliary heating/ cooling systems

A major objective of solar building design is the reduction or, if possible, elimination of the need for auxiliary heating or cooling. By utilizing the thermal storage capacity of the building, the designer may reduce the size of auxiliary heating/cooling systems. Heating systems may generally be separated into two main categories: convective heating systems that heat primarily the room air, and radiant or panel heating systems in which the source of heat is integrated into the building envelope or interior surfaces. Radiant heating systems operate in a mode similar to direct gain systems, that is, they rely on a high surface temperature (up to 29°C for floor heating) in a room, and lower air temperature.

Section 7.1 considers a convective system while Section 7.2 focuses on a floor heating system. Section 7.3 describes the analysis of a ground cooling/heating system integrated with central air-conditioning.

7.1 Heating and cooling load calculations

7.1.1 Mathematical model

Calculation of peak heating and cooling loads should take into account building thermal storage capacity and dynamic variation of solar radiation and outdoor temperature in order to avoid oversizing of auxiliary heating and cooling systems. For most mild temperate climates, a heat pump will provide an efficient auxiliary heating and cooling system. Well-insulated buildings with effective shading systems and natural ventilation will minimize the need for auxiliary cooling. Similarly, appropriate sizing of the fenestration systems facing towards the equator (south or near-south in the northern hemisphere) will satisfy most heating requirements on sufficiently clear days.

The design of a building auxiliary heating/cooling system based on passive solar principles should therefore be based on the following principles:

- The heating system should be sized based on a sequence of cloudy cold days, but with the thermal mass utilized to minimize peak loads.

- The cooling system capacity should be estimated based on a sunny hot day but with the use of natural ventilation and effective shading devices.
- A variable room thermostat setpoint should be employed when necessary to reduce peak loads.
- If a convective heating/cooling system such as a heat pump is utilized, the air distribution system should be designed for acceptable thermal comfort.

Analysis is performed for a design day with major objective to determine peak heating or cooling load.

The principle of mass utilization to reduce peak heating or cooling loads is simple. In the case of heating, the peak demand is expected to occur just before sunrise at about 6 a.m. If the thermal storage mass of the building is kept relatively warm at night, then the peak load will be reduced. Similarly, on a hot day, the peak load will typically occur in the afternoon; if the building mass is precooled in the morning or at night by natural ventilation, then the peak cooling demand will be reduced.

Traditionally, heating systems for traditional buildings were oversized by using a steady state peak heat loss calculation technique. For peak cooling load calculations, a steady periodic method is often employed such as the CLTD (cooling load temperature difference) recommended by ASHRAE (1997). Here, it is proposed to employ a steady-periodic admittance method, similar to the one employed for calculation of temperature swings.

A simple analytical solution for the heating/cooling source may be obtained by specifying the zone temperature and assuming combined radiative-convective surface conductances between the heat conductive elements (e.g. walls) and the room air. We thus have a star thermal network and parallel heat transfer paths between inside and outside similar to the calculation of the house/zone steady state total conductance as shown in Figure 7.1. For steady periodic calculations, a frequency domain technique may be employed to calculate the zone admittance for about three frequencies plus the mean (steady state). By utilizing complex number algebra and Fourier series, we may then compute design day profiles for auxiliary heating (positive) and cooling (negative) loads.

The main advantage of the admittance method is that it is convenient for steady-periodic analysis (with about three harmonics). However, it cannot accommodate non-linearities. The explicit finite difference method may accommodate non-linear heat transfer coefficients, and control strategies. Here, we will consider a detailed model based on the **admittance method**. Each wall is modelled by a self-admittance and a transfer admittance as described in Chapter 5. The following inputs are required.

The **energy balance** may be written in a similar form to that for the simple model of equation (5.1) used for the passive analysis. All walls are represented by their self-admittances, together with the equivalent heat source Q_{sc} at their inside surface, equal to the transfer admittance times the outside or other equivalent

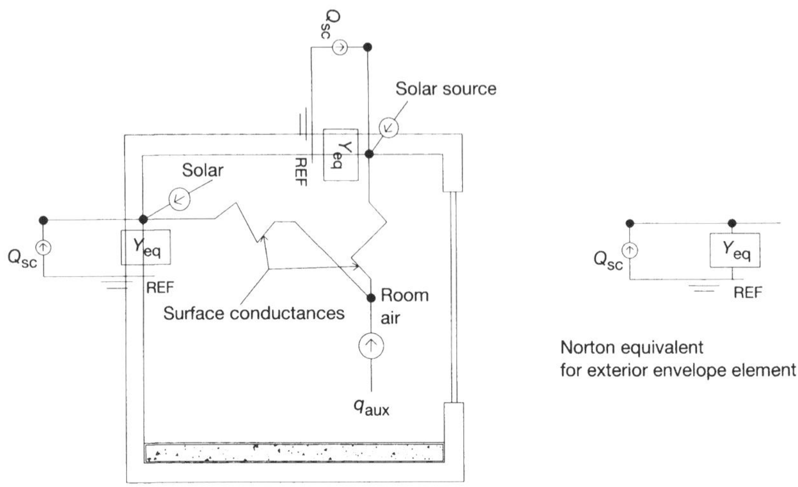

Figure 7.1 Schematic of zone thermal network showing Norton equivalent for typical walls; Y_{eq} is admittance and Q_{sc} is equivalent heat source

known temperature as shown in Figure 7.1. Radiation and convection in the room are represented by combined film coefficients. Thus, an analytical solution for room temperature or auxiliary heating may be obtained. A more detailed model with separate radiation exchanges requires an energy balance at the interior surfaces (Athienitis *et al.* 1990).

Energy balance at room air node (in admittance form):

$$[Y]\{T\} = \{Q\} \tag{7.1a}$$

where $[Y]$ is the admittance matrix and $\{T\}$ and $\{Q\}$ the temperature and source vectors respectively.

Since we have a star network at the room air node (1), all heat sources at surfaces may be transformed to equivalent heat sources at the air node by multiplying by the factor $U_i/(U_i + Y_s)$ where U_i is the surface conductance ($A \times h_i$) and Y_s the self-admittance of each wall. Thus, surface nodes may be eliminated and we obtain the following energy balance at the room air node:

$$Y_z T_R = \Sigma Q_{eq} + q_{aux} \tag{7.1b}$$

where q_{aux} = auxiliary heating (or cooling if negative), Y_z is room admittance and T_R is room air temperature.

All heat gains/losses are expressed as equivalent heat sources (negative if losses) Q_{eq} (the short-circuit source Q_{sc} in Figure 7.1 is similar to Q_{eq}) at the room temperature node based on a reference temperature of 0°C as follows:

1. Internal gains plus heat source due to infiltration

$$= Q_{\text{intc}_n} + \left(U_{\text{inf}} + \sum_{iw} U_{\text{w}_{iw}} \right) T_{\text{o}_n} \tag{7.2}$$

2. Equivalent heat sources at wall interior surfaces due to sol-air temperatures or specified temperatures in adjacent zones. These can be added to internal radiant gains Q_{intr} at the interior surfaces:

$$\left(Y_{\text{t}_{n,i}} \cdot T_{\text{eq}_{n,i}} + Q_{\text{intr}_{n,i}} \right) \tag{7.3}$$

The source in equation 7.3 must be converted to a heat gain at the room air point so that we can perform an energy balance at this node. The equivalent heat source at the air node is given by:

$$\left(Y_{\text{t}_{n,i}} \cdot T_{\text{eq}_{n,i}} + Q_{\text{intr}_{n,i}} \right) \cdot \frac{U_{\text{i}_i}}{Y_{\text{s}_{n,i}} + U_{\text{i}_i}} \tag{7.4}$$

The energy balance to determine auxiliary heating or cooling q_{aux} yields:

$$q_{\text{aux}_n} = \left[\text{TR}_n - \frac{\left[Q_{\text{int}_n} + \left(U_{\text{inf}} + \sum_{iw} U_{\text{w}_{iw}} \right) \cdot T_{\text{o}_n} \right] + \sum_i \left[\left(Y_{\text{t}_{n,i}} \cdot T_{\text{eq}_{n,i}} + Q_{\text{intr}_{n,i}} \right) \cdot \frac{U_{\text{i}_i}}{Y_{\text{s}_{n,i}} + U_{\text{i}_i}} \right]}{Y_{\text{z}_n}} \right] \cdot Y_{\text{z}_n} \tag{7.5}$$

Equation (7.5) is valid for the steady state (mean) as well as for all harmonics n, a consequence of the superposition principle for linear systems. It is equally valid for heating and cooling analysis. In all cases, the auxiliary heating/cooling convective source is assumed to act at the air node (representing the air in the zone), which is kept at a specified temperature $T_{\text{R}}(t)$ which could be constant, square-wave, sinusoidal or any other desired function.

For heating load calculations we will consider a minimal level of solar radiation equivalent to a cloudy day. The outside temperature will be modelled by a three-harmonic Fourier series based on its value at eight different times.

The model employed for cooling load calculations is the same as the model used for heating load calculations. The main difference here is that solar radiation transmitted through windows or absorbed by outside surfaces is calculated in detail. Also, latent cooling load is calculated. A passive analysis may be performed by setting the source q_{aux} to 0 in equation 7.5 and solving for the room temperature.

The following weather inputs and building data are required for the calculation.

WEATHER INPUTS

- Outside temperature (daily mean and range).
- Solar radiation transmitted by each window and solar radiation absorbed by each exterior surface. For a clear day (cooling load analysis or passive solar analysis) this can be readily estimated from Hottel's clear sky model. For heating loads, a conservative low estimate of solar radiation may be used.

BUILDING DATA

NS = number of surfaces contributing to the zone energy balance.
NS_e = number of exterior surfaces (walls and roof).
A_i = area of exterior surface i.
N_w = number of windows; A_{wi} = area of window i.
ψ_{se} with se = $0,1..N_{se}$ exterior surface azimuth angle.
α_{se} with se = $0,1..N_{se}$ solar absorptance of exterior surfaces.
Window type: U-value or thermal resistance, single or double glazing and kL value (extinction coefficient × thickness) – for cooling load.
A_{door} = external door area; R_{door} = external door R-value.
Wall construction: Wall layer properties. Interior layer properties for transient analysis are also required.
ach: infiltration – air changes per hour.
h_i: inside surface heat transfer coefficient for surface i (typical combined radiative convective coefficients are assumed).
Internal gains: Q_{intr} = radiative internal gains; Q_{intc} = convective internal gains

Example

Consider a house that consists of a basement and a ground-level floor, with a pitched roof (Figure 7.2). Here we consider the ground-level zone.

H_h = 2.7 m, L_h = 14 m, W_h = 12 m
Surfaces contributing to energy balance:
NS = 6 (= 1,2..NS)
1–4 walls, 5-ceiling, 6-floor
se = 1..5 exterior surfaces
N_w = 4 (number of windows), i = $1,2..N_w$
(assume four windows – sum the window areas on each house side)
$\psi_1 = -45°, \psi_2 = -135°, \psi_3 = 135°, \psi_4 = 45°$
αs_{iw} = 0.9 wall absorptances
Window and door areas: A_{w1} = 12 m², A_{w2} = 3 m², A_{w3} = 2 m², A_{w4} = 3 m²
A_{d1} = 2 m², A_{d2} = 2 m², A_{d3} = 2 m², A_{d4} = 2 m²

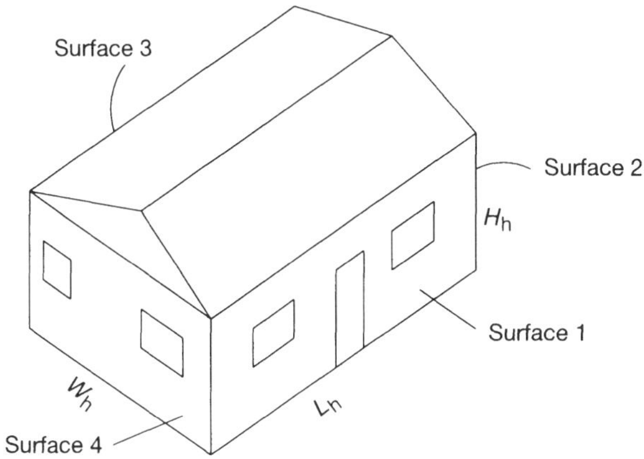

Figure 7.2 Schematic of one-zone house considered in Example

Wall net areas:

$$A_1 = L_h H_h - A_{w1} - A_{d1}$$

$$A_2 = W_h H_h - A_{w2} - A_{d2}$$

$$A_3 = L_h H_h - A_{w3} - A_{d3}$$

$$A_4 = W_h H_h - A_{w4} - A_{d4}$$

$$A_5 = W_h L_h - A_{w5} - A_{d5}$$

Internal height, $H_I = 2.4$ m
Volume $= A_5 H_i$
Door thermal resistance, $R_d = 1$ m²°C/W
Window resistance (double glazed), $R_w = 0.34$ m²°C/W
Film coefficients, $h_1 = 8.3$ W/m²°C, $h_2 = h_1, h_3 = h_1, h_4 = h_1$
$h_5 = 9.3$ W/m²°C, $h_6 = 6.1$ W/m²°C (cold floor)
air changes/hour, ach $= 0.5$
Calculation of infiltration conductance (infiltration heat transfer):
Specific heat, $c_{pair} = 1000$ J/kg°C
Density of air, $\rho_{air} = 1.2$ kg/m³
$U_{inf} = (\text{ach} \times \text{Vol})/(3600.\ \text{s}) \times \rho_{air} \times c_{pair}$
$U_{inf} = 67.2$ W/°C

THERMAL RESISTANCE OF WALLS (INCLUDING AIR FILMS)

To calculate the effective resistance of a building envelope element, we typically assume parallel heat flow paths through each area with a different composition (e.g. through wood structure versus through insulation).

Vertical walls (compute resistance assuming parallel heat flow paths)
Gypsum board

Thickness, L_1 = 0.013 m
Density, ρ_1 = 800 kg/m^3
Conductivity, k_1 = 0.16 W/m°C
Specific heat, c_1 = 750 J/kg°C

Insulation

R_{ins} = 2.2 m^2°C/W

Siding + sheathing

R_{sid} = 0.37 m^2°C/W

Exterior film

h_o = 22 W/m^2°C (assume)

Assume 15% of wall area is structural framing. The fraction of area that is framing:

f_f = 0.15

2 by 4 wood stud with R-value (with framing resistance)

R_f = 0.77 m^2°C/W

$$R_1 = \frac{1}{\dfrac{1-f_f}{\left(\dfrac{L_1}{k_1}+R_{ins}+R_{sid}+\dfrac{1}{h_o}+\dfrac{1}{h_1}\right)}+\dfrac{f_f}{\left(\dfrac{L_1}{k_1}+R_f+R_{sid}+\dfrac{1}{h_o}+\dfrac{1}{h_1}\right)}}$$

$$= 2.44\,°\text{C}\,\text{m}^2\,/\,\text{W}$$

Calculation of wall conductance excluding interior layer and film (to be used for admittance calculations):

$$u_1 = \frac{1}{R_1 - \dfrac{L_1}{k_1} - \dfrac{1}{h_1}}$$

Assume that all exterior walls (ii) are of the same construction:
$ii = 1,2..4, L_{ii} = L_1, R_{ii} = R_1$
$u_{ii} = u_1, k_{ii} = k_1, \rho_{ii} = \rho_1, c_{ii} = c_1$

Calculation of roof-ceiling thermal resistance

Ceiling

Gypsum board

$L_5 = L_1, k_5 = k_1, c_5 = c_1, \rho_5 = \rho_1$

Insulation

$R_{insc} = 2.8 \ \mathrm{m^2{}^\circ C/W}$

Air-film (attic)

$h_a = 12 \ \mathrm{W/m^2{}^\circ C}$

$$R_c = \frac{1}{\dfrac{1-f_f}{\left(\dfrac{L_5}{k_5} + R_{insc} + \dfrac{1}{h_a} + \dfrac{1}{h_5}\right)} + \dfrac{f_f}{\left(\dfrac{L_5}{k_5} + R_f + \dfrac{1}{h_a} + \dfrac{1}{h_5}\right)}}$$
$$= 2.377 ^\circ\mathrm{C\,m^2/W}$$

$R_c = 2.377 \ ^\circ\mathrm{C\ m^2/W}$

Roof

Exterior air film

$h_o = 20 \ \mathrm{W/m^2{}^\circ C}$

Shingle backer board

$R_b = 0.19 \ \mathrm{m^2{}^\circ C/W}$

Wood shingles

$R_{sh} = 0.17 \ \mathrm{m^2{}^\circ C/W}$

$$R_r = \frac{1}{\dfrac{1-f_f}{\left(R_b + R_{sh} + \dfrac{1}{h_o} + \dfrac{1}{h_a}\right)} + \dfrac{f_f}{\left(R_f + R_b + R_{sh} + \dfrac{1}{h_o} + \dfrac{1}{h_a}\right)}}$$
$$= 0.53 ^\circ\mathrm{C\,m^2/W}$$

Assuming a 30° slope for the roof, we calculate the ceiling-roof combined resistance per unit ceiling area as follows:

$A_r = A_5/\cos\ 30°$

$R_5 = (R_c/A_5) + (R_r/A_r) \times A_5$

For admittance calculation:

$u_5 = 1/[R_5 - (L_5/k_5) - (1/h_5)]$

$R_5 = 2.848\ \text{m}^2\text{°C/W}$

Floor

Carpet and underpad

$L_6 = 0.02\ \text{m}, k_6 = 0.06\ \text{W/m°C}, \rho_6 = 800\ \text{kg/m}^3$

Insulation and plywood

$R_{ins} = 1.0\ \text{m}^2\text{°C/W}, c_6 = 1400\ \text{J/kg°C}$

Air film (horizontal heat flow downward)

$h_o = 6.13\ \text{W/m}^2\text{°C}, R_6 = R_{ins} + (L_6/k_6) + (1/h_o) + (1/h_6),$
$R_6 = 1.66\ \text{m}^2\text{°C/W}, u_6 = 1/[R_6 - (L_6/k_6) - (1/h_6)]$

CALCULATION OF WALL ADMITTANCES

The self-admittance and the transfer admittance will be calculated for each wall, considering the thermal capacity of the room interior layer. Note that the steady-state value of the admittance is equal to the wall conductance. We will calculate admittances from outside to the interior surface and to the room air point. The analysis will be performed for the mean term and three harmonics of the weather inputs.

Admittances

Steady-state admittance is equal to wall *U*-value (excluding interior film); first subscript indicates frequency, second subscript indicates surface number.

$$Y_{s_{0,i}} = \frac{A_i}{R_i - \dfrac{1}{h_i}}$$

Admittances from outside to room air (steady state):

$$Y_{t_{0,i}} = Y_{s_{0,i}}$$

$$Y_{0,i} = \frac{A_i}{R_i} \qquad Y_{ta_{0,i}} = Y_{0,i}$$

$N_h = 5, n = 1, 2 \ldots N_h$

Let $j = \sqrt{-1}$

$$\gamma_{n,i} = \sqrt{j\frac{2\pi n}{\frac{k_i}{\rho_i c_i} \times 86400\,\text{s}}}$$

Interior and exterior surface conductances:

$$U_{i_i} = A_i h_i \quad U_{o_i} = h_o A_i$$

$$I_w = 1,2..4$$

Total conductance of double-glazed windows and doors:

$$U_{w_{iw}} = \frac{A_{w_{iw}}}{R_w} + \frac{A_{d_{iw}}}{R_d}$$

Self-admittance Y_s and transfer admittance Y_t for each wall:

$$Y_{s_{n,i}} = A_i \frac{u_i + k_i \gamma_{n,i} \tanh(\gamma_{n,i} L_i)}{\left(\frac{u_i}{k_i \gamma_{n,i}} \tanh(\gamma_{n,i} L_i)\right) + 1}$$

$$Y_{t_{n,i}} = \frac{A_i}{\frac{\cosh(\gamma_{n,i} L_i)}{u_i} + \frac{\sinh(\gamma_{n,i} L_i)}{k_i \gamma_{n,i}}}$$

Wall admittances from outside to inside air:

$$Y_{n,i} = \frac{Y_{s_{n,i}} U_{i_i}}{Y_{s_{n,i}} + U_{i_i}}$$

$$Y_{ta_{n,i}} = Y_{t_{n,i}} \frac{U_{i_i}}{Y_{s_{n,i}} + U_{i_i}}$$

Zone admittance Y_z (from air point):

$$Y_{z_n} = U_{\text{inf}} + \sum_{iw} U_{w_{iw}} + \sum_{i} Y_{n,i}$$

Note that Y_{z0} is simply the total U-value of the house. The value of Y_z for the steady state and first three harmonics is given by:

$$Y_z = \begin{bmatrix} 340.267 \\ 403 + 298.766j \\ 539.342 + 519.654j \\ 683.104 + 670.462j \end{bmatrix} \text{W / °C}$$

Outside temperature (T_o)

The outside temperature is modelled by a Fourier series based on N_{To} + 1 values. If more detail is required, N_{To} may be increased. In this example assume 3 hour intervals.

Time index (*it*):

$$N_{To} = 7,\ it = 0,\ 1..N_{To}$$

Time:

$$t_{it} = \text{it}\ 3\ \text{h}$$

Harmonics:

$$n = 0,1..N_h$$

$$w_n = 2\pi n/24\ \text{h}$$

The harmonics of T_o are calculated from the NT_o values with the following discrete Fourier transform (DFT):

$$T_{on_n} = \left(\sum_{it} T_{o_{it}} \frac{\exp(-j w_n \cdot t_{it})}{N_{To} + 1} \right) {}^\circ C$$

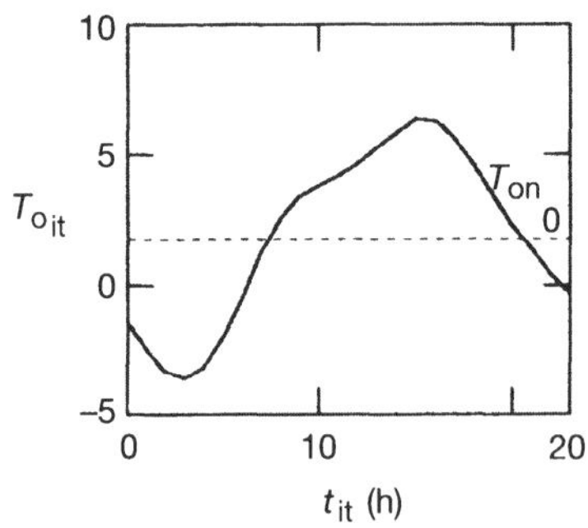

Figure 7.3 Outdoor temperature (from hourly or 3-hour interval values)

Harmonics of T_o:

$$T_{on} = \begin{bmatrix} 1.875 \\ -1.811 + 1.332j \\ -0.125 + 0.25j \\ 0.311 + 0.082j \end{bmatrix}$$

Mean daily temperature:

$$T_{on,0} = 1.875°C$$

The T_o curve in Figure 7.3 was calculated from the following inverse discrete Fourier transform:

$$T_{o_{it}} = T_{on_0} + 2\sum_{n1} \mathrm{Re}\left[\left(T_{on_{n1}}\right)\exp\left(j\,w_{n1} \cdot t_{it}\right)\right]$$

where $n1 = 1,2,3$

SOLAR RADIATION

Solar radiation will be modelled by a half-sinusoid from sunrise ($-t_s$)to sunset time t_s.

Redefine time array:

$it = 0,1..23$, $t_{it} = it$ h

Let $t_s = 5$ h (time from solar noon to sunset). The peak solar radiation at noon (assuming minimal level for a cloudy day) will be:

$$S_{max} = 100 \text{ W/m}^2$$

$$f_{it} - S_{max} \cos\left(\pi \cdot \frac{t_{it} - 12\text{h}}{2t_s}\right)$$

(positive function, otherwise set to 0 – at night)

$S(t)$ may be modelled with a discrete Fourier series as follows. We multiply by total window area to determine total solar radiation:

$$S_{n_n} = \left(\sum_{it} S_{it} \frac{\exp\left(-j w_n t_{it}\right)}{24}\right) \sum_{iw} A_{w_{iw}}$$

We will assume that the fraction of this radiation absorbed by each interior surface is proportional to its area:

$$A_{tot} = \Sigma_i A_i$$

$$Q_{r_{n,i}} = S_{n_n} \frac{A_i}{A_{tot}}$$

We may similarly model internal gains with Fourier series. In this example, we will ignore internal gains. Also, we will ignore solar radiation absorbed on wall exterior surfaces since our **objective is to determine the peak heating load**.

Room-air temperature T_{R} (assume specified):

$$T_{\mathrm{R}_0} = 20°\mathrm{C} \qquad n1 = 1,2..3 \qquad T_{\mathrm{R}_{n1}} = 0°\mathrm{C}$$

A variable room temperature may be employed, e.g. with a night setback/setup. In such a case T_{R} should be modelled by a Fourier series like T_{o}.

Equivalent temperatures for exterior surfaces (ignoring solar effects):

$$T_{\mathrm{eq}_{n,se}} = T_{\mathrm{on}_n}$$

For basement:

$$T_{\mathrm{b}} = 20°\mathrm{C} \qquad T_{\mathrm{eq}_{0,6}} = T_{\mathrm{b}} \qquad T_{\mathrm{eq}_{n1,6}} = 0°\mathrm{C}$$

First determine the mean auxiliary load:

$$q_{\mathrm{aux}_0} = \left[T_{\mathrm{R}_0} - \frac{\left(U_{\mathrm{inf}} + \sum_{iw} U_{\mathrm{w}_{iw}} \right) T_{\mathrm{on}_0} + \sum_i \left(Y_{t_{0,i}} \cdot T_{\mathrm{eq}_{0,i}} + Q_{\mathrm{r}_{0,i}} \right) \cdot \frac{U_{\mathrm{i}_i}}{Y_{\mathrm{s}_{0,i}} + U_{\mathrm{i}_i}}}{Y_{\mathrm{z}_0}} \right] \cdot Y_{\mathrm{z}_0}$$

$$= 3.841 \times 10^3\,\mathrm{W}$$

$$q_{\mathrm{aux}_{n1}} = \left[T_{\mathrm{R}_{n1}} - \frac{\left(U_{\mathrm{inf}} + \sum_{iw} U_{\mathrm{w}_{iw}} \right) T_{\mathrm{on}_{n1}} + \sum_i \left(Y_{t_{n1,i}} \cdot T_{\mathrm{eq}_{n1,i}} + Q_{\mathrm{r}_{n1,i}} \right) \cdot \frac{U_{\mathrm{i}_i}}{Y_{\mathrm{s}_{n1,i}} + U_{\mathrm{i}_i}}}{Y_{\mathrm{z}_{n1}}} \right] \cdot Y_{\mathrm{z}_{n1}}$$

Auxiliary heating power. This is calculated using an inverse discrete Fourier series:

$$q_{\mathrm{auxt}_{it}} = q_{\mathrm{aux}_0} + 2\sum_{n1} \mathrm{Re}\left[q_{\mathrm{aux}_{n1}} \exp(j\, w_{n1} t_{it}) \right]$$

Peak heating load (used to size heating system):

$$\max(q_{\mathrm{auxt}}) = 5.641 \times 10^3\ \mathrm{W}$$

Heating energy consumption Q_{h} obtained by numerical integration of the positive part of the auxiliary power curve:

$$Qh = \sum_{it} \frac{q_{\text{aux}_{it}} + \left| q_{\text{auxt}_{it}} \right| + q_{\text{auxt}_{it+1}} + \left| q_{\text{auxt}_{it+1}} \right|}{4} \left(t_{it+1} - t_{it} \right)$$

$$= 331.838 \text{ MJ}$$

which is equal to

$$q_{\text{aux}_0} 24\,\text{h} = 331.838 \text{ MJ}$$

Note that in this case, because there is no cooling, Q_h may be obtained from the mean power.

A variable room temperature may be employed, e.g. with a night setback. In such a case, T_R should be modelled as a Fourier series.

Note that combined radiative-convective coefficients were employed in the above model in order to permit a simple analytical solution. A more detailed model is described by Athienitis *et al.* (1990) and utilized in Chapter 8. The method is similar in concept to the CLTD and transfer functions methods of ASHRAE (1997). However, all building transfer functions (admittances) are directly calculated.

Next, we will consider the same analysis but with a night setback in the thermostat setpoint (Figure 7.4).

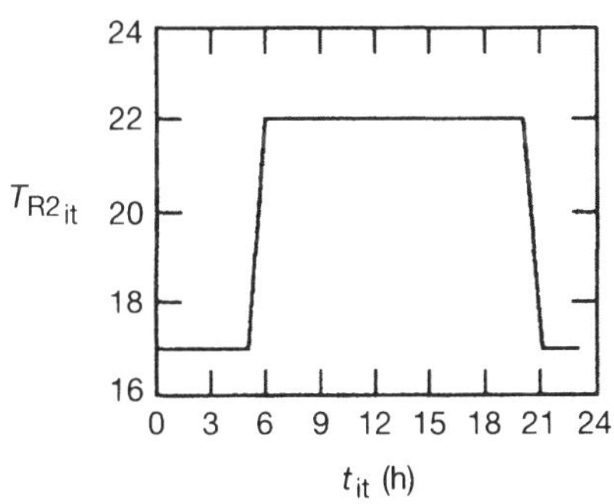

Figure 7.4 Thermostat night setback setpoint profile

First represent the thermostat setpoint with a discrete Fourier series whose coefficients are:

$$T_{R2n_n} = \left(\sum_{it} T_{R2_{it}} \frac{\exp(-j w_n t_{it})}{24} \right) {}^\circ\text{C}$$

Then determine the mean auxiliary load:

$$q^2{}_{\text{aux}_0} = \left[T_{\text{R2n}_0} - \frac{\left(U_{\text{inf}} + \sum_{iw} U_{\text{w}_{iw}} \right) T_{\text{on}_0} + \sum_i \left(Y_{t_{0,i}} \cdot T_{\text{eq}_{0,i}} + Q_{\text{r}_{0,i}} \right) \cdot \frac{U_{\text{i}_i}}{Y_{\text{s}_{0,i}} + U_{\text{i}_i}}}{Y_{\text{z}_0}} \right] \cdot Y_{\text{z}_0}$$

$$= 3.883 \times 10^3 \,\text{W}$$

(close to that without the night setback since the average setpoint is approximately the same).

$$q^2{}_{\text{aux}_{n1}} = \left[T_{\text{R2n}_{n1}} - \frac{\left(U_{\text{inf}} + \sum_{iw} U_{\text{w}_{iw}} \right) T_{\text{on}_{n1}} + \sum_i \left(Y_{t_{n1,i}} \cdot T_{\text{eq}_{n1,i}} + Q_{\text{r}_{n1,i}} \right) \cdot \frac{U_{\text{i}_i}}{Y_{\text{s}_{n1,i}} + U_{\text{i}_i}}}{Y_{\text{z}_{n1}}} \right] \cdot Y_{\text{z}_{n1}}$$

Auxiliary power with night setback is given by:

$$q^2{}_{\text{auxt}_{it}} = q^2{}_{\text{aux}_0} + 2 \sum_{n1} \text{Re}\left[q^2{}_{\text{aux}_{n1}} \exp\left(j\, w_{n1} t_{it} \right) \right]$$

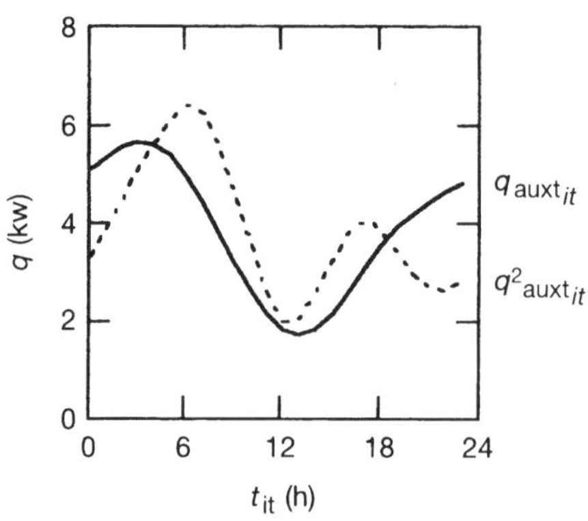

Figure 7.5 Auxiliary heating load with constant setpoint (q_{auxt}) and with setback ($q^2{}_{\text{auxt}}$)

Peak heating load (used to size heating system):

$$\max\,(q2_{\text{auxt}}) = 6.3252 \times 10^3 \;\text{W}$$

We therefore conclude that the peak heating load increases owing to the sudden increase in setpoint (from 17°C to 22°C) even though the mean load is approximately the same (Figure 7.5).

An effective strategy to reduce energy consumption at night while avoiding high peak loads is to have a slowly changing setpoint in the early morning such as a sinusoidal or ramp change.

7.1.2 Special considerations for cooling

For peak cooling load estimates to size cooling systems the following principles should be kept in mind:

- The best method to lower sensible cooling loads is to prevent solar radiation from being transmitted into the building interior through effective, preferably exterior, shading devices. All building envelope elements should also have effective insulation and windows should be at least double-glazed.
- The solar radiation may be modelled in more detail as indicated in Chapter 3.
- Sensible cooling loads may be lowered through night-time natural ventilation, and evaporative cooling.
- The same basic thermal model as for heating loads may be employed, with the inclusion of internal gains, transmitted solar radiation and for the exterior walls, sol-air temperatures.

For the previous example we obtain the solar radiation curves given in Figures 7.6 and 7.7 (latitude 45°N). Without shading, a peak load of about 11 kW is obtained. When shading is employed, the peak sensible load is reduced by about 50%.

The **latent cooling load** is due to exchange of inside air and outside air with different moisture contents and due to generation of moisture by occupants, equipment, etc. If moisture must be removed from the outdoor air to reduce the humidity level to the comfort range, the energy required (assume indoor relative humidity 50% and outdoor humidity 80%) is determined as follows.

Latent heat of water:

$$h_{\mathrm{fg}} = 2465 \ \mathrm{kJ/kg}$$

Interior and exterior humidity ratios:

$$W_{\mathrm{i}} = 0.0073$$

$$W_{\mathrm{o}} = 0.0117$$

$$q_{\mathrm{lat}} = (\mathrm{ach}/3600)\rho_{\mathrm{air}}\mathrm{Vol}(W_{\mathrm{o}} - W_{\mathrm{i}})h_{\mathrm{fg}} = 728 \ \mathrm{W}$$

Note that there is also a small, usually negligible, additional load (sensible) due to the difference in temperature between the room air moisture and the outside air moisture.

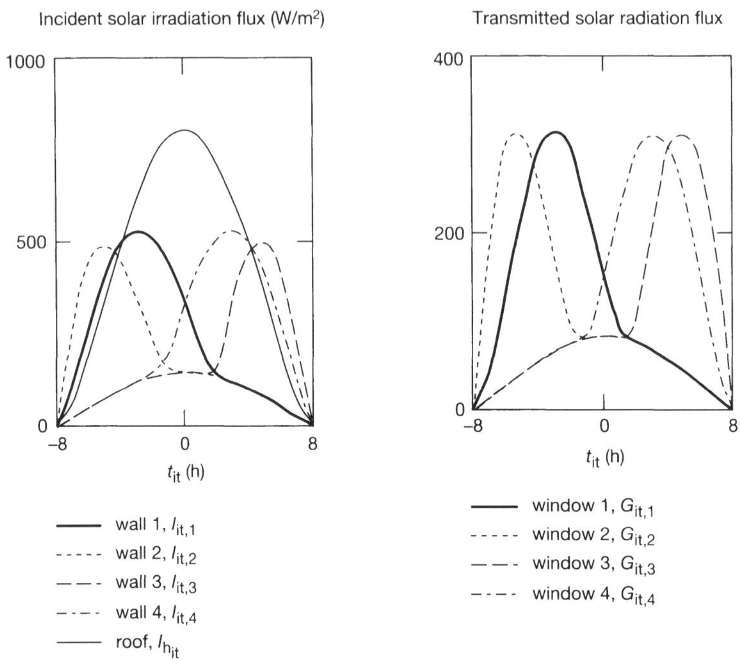

Figure 7.6 Solar irradiance on exterior surfaces I and transmitted irradiation G (assuming no shading) based on Hottel's clear sky model (see Chapter 3)

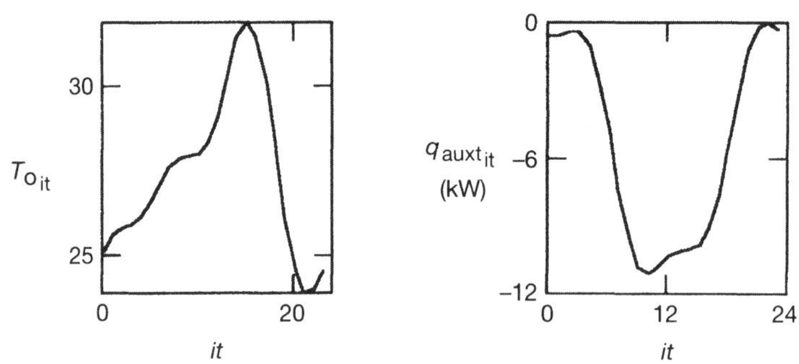

Figure 7.7 Outside temperature profile employed for cooling load calculation

An internal heat gain of approximately 67 W per person (for residences) may also be included, as well as a heat gain of 470 W from appliances. These heat gains may be assumed to be approximately 50% convective and 50% radiative (absorbed by each wall proportionally to its area).

7.2 Analysis and design of solar-assisted radiant heating systems

7.2.1 Introduction

This section presents a model of an integrated radiant floor heating–direct gain passive solar system. Thermal mass is utilized for both storage of auxiliary heating energy and direct solar gains incident on the floor. An explicit finite difference model is employed to study the performance of a passive solar building.

A direct-gain passive solar building usually combines efficient south-facing fenestration systems which transmit high amounts of solar radiation, while reducing heat losses to the outdoors, with an optimal amount of thermal storage mass. The thermal storage mass should be distributed on an interior building surface directly illuminated with transmitted solar radiation. Also, sufficient quantities of thermal mass must be employed to store passive solar gains while reducing temperature swings. Optimal quantities of thermal mass for storage of passive solar gains so as to reduce room temperature swings may be determined with frequency domain techniques (see Chapter 5); however, this analysis does not consider the effect of heating system type.

A floor heating system is potentially highly compatible in operation with a direct gain passive solar system which would also store solar gains within the floor mass. However, there is a significant difference between the solar gains and the auxiliary heat applied; the solar gains are absorbed at the top surface of the floor slab, while the auxiliary heat is supplied at the bottom of the slab. The slab introduces a different thermal lag effect in each case.

Design of floor heating systems requires several important decisions on a number of design or control variables as follows:

- Maximum heating device output (maximum auxiliary heating).
- Type and thickness of thermal mass integrated in the floor heating system.
- Appropriate control strategies for the system to maintain desired thermal conditions in the heated space and to prevent the floor surface temperature from exceeding 29°C as required by current standards (ASHRAE 1987).

All three design decisions are interrelated and should be considered simultaneously in order to achieve energy-efficient operation of the system and satisfactory thermal comfort. There are various design approaches that may be followed. Traditionally, high amounts of thermal mass were used (10 cm thick concrete or equivalent) with continuous heating and a fairly constant indoor temperature as well as a relatively low floor surface temperature (24–26°C). Recently, with the increased use of floor heating systems in timber frame houses, lower amounts of thermal mass are encountered – often as little as a 3 cm thick concrete slab or equivalent.

Peak loads may also be reduced and shifted to night-time when electricity costs may be lower. The key problem with the performance of a passive solar building with a floor heating system is overheating. This is because when a significant amount of solar radiation is incident on the floor, its thermal storage mass is already warm from continuous auxiliary heating input during the night and in the early morning. Also, the floor surface temperature may in some cases exceed the comfort limit. These problems may be addressed by lowering the setpoint at night and slowly increasing it during the daytime, or alternatively increasing the amount of thermal storage mass. Here, we focus on the first option, which essentially involves anticipatory or predictive control. Therefore, such systems may present comfort control problems and lead to overheating. These problems may be reduced by employing various forms of anticipatory control such as ambient temperature reset of setpoints, with which the indoor setpoint or the instantaneous auxiliary heat energy output is linked to the outdoor temperature. Usually, the reset ratio relating outside temperature to supply water temperature in hydronic systems must be adjusted after installation to take into account the as-built system characteristics. However, this frequently used technology for control of radiant heating systems does not solve the current problem because the outside temperature usually increases significantly in the afternoon after most of the solar gains have been absorbed.

Proportional control of room temperature with a constant setpoint is effective. However, when energy-saving strategies are implemented, such as storage of passive solar gains in the floor and a lower setpoint at night, effective control becomes difficult. For example, when a simple thermostat setpoint night setback is employed, it contributes to energy conservation but leads to a very high increase in the peak heating load on cold days. If this demand is not satisfied by using a large-capacity heating system, the room temperature will rise too slowly to the desired temperature. Thus, the setpoint should be gradually changed, starting a few hours before the desired effect. If storage of solar gains is to be effectively achieved, then a lower night-time setpoint is required to keep the floor mass cool in anticipation of the upcoming solar gains, followed by a gradual increase in the morning to the desired comfort range.

7.2.2 Mathematical model

The explicit finite difference method is particularly suitable for modelling of non-linear heat diffusion problems such as the present case of heat transfer through the floor heating system and its control. It can easily accommodate non-linear heat transfer coefficients and temperature-dependent heat sources, such as auxiliary heating; non-linear simulations were performed in the present study.

Figure 7.8 shows the thermal network for the zone and floor heating system. The finite difference thermal network model for the room consists of two capacitances for the floor thermal capacity, interconnecting thermal resistances, and another thermal capacitance for the rest of the room. Node 1 represents room air. Node 2 is the floor surface and node 6 the unheated surfaces; both surfaces are coupled by convection with room air (resistances R_{12} and R_{16}) and together by radiation (R_{26}). The energy balance for the thermal network is:

$$T_{i,p+1} = \frac{\Delta t}{C_i}\left(q_i + \sum \frac{T_{j,p} - T_{i,p}}{R_{i,j}}\right) + T_{i,p} \tag{7.6}$$

Subscript i indicates the node for which the energy balance is written and j all nodes connected to node i, while p is the time interval; q_i represents a heat source at node i such as auxiliary heating (node 4) or solar radiation (nodes 2 and 6).

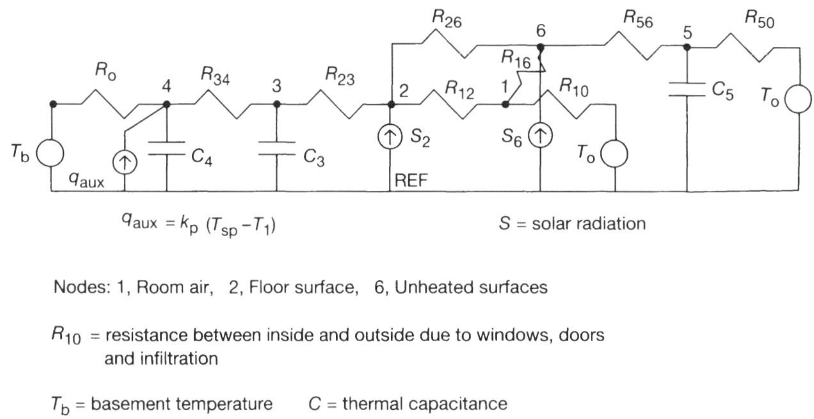

Figure 7.8 Thermal network of zone with floor heating system (corresponding to house of Figure 7.2)

The time step is selected based on the following condition for numerical stability (Holman 1986):

$$\Delta t \leq C_i \Sigma_j (1/R_{i,j}) \tag{7.7}$$

for all nodes i.

For the room under consideration, the time step selected was 400 seconds. The solar radiation transmitted by the windows is assumed to be absorbed 70% by the floor and 30% by the remaining surfaces.

This section presents a model for a space (one zone) heated with a floor heating system. It is assumed that the heat delivered by the system is q_{aux}.

The auxiliary heating q_{aux} is proportional to the error between the setpoint T_{sp} and the actual room air temperature T_1:

$$q_{aux} = K_p(T_{sp} - T_1) \tag{7.8}$$

where K_p = proportional control constant

Note that the source of the heat could be electric panels or hot water pipes covered by a layer of thermal storage mass such as light concrete plus ceramic tiles. If we have a hydronic hot water system, then the heating power delivered is also equal to the mass flow rate times the difference in temperature between supply and return.

Example

Consider a house as in the previous section, which consists of a basement and a ground-level floor, in Montreal, Canada. Here we consider the ground-level zone, heated with floor heating.

$H_h = 3$ m, $L_h = 15$ m, $W_h = 10$ m, $N_w = 4$, $iw = 1,2..N_w$
(assume four windows – sum the window areas on each house side)
Window and door areas:

$A_{w1} = 8\ \text{m}^2, A_{w2} = 4\ \text{m}^2, A_{w3} = 4\ \text{m}^2, A_{w4} = 4\ \text{m}^2$
$A_{d1} = 1.8\ \text{m}^2, A_{d2} = 1.8\ \text{m}^2, A_{d3} = 1.8\ \text{m}^2, A_{d4} = 1.8\ \text{m}^2$

Door thermal resistance:

$$R_d = 2\ \text{m}^2\text{°C/W}$$

Window resistance (double glazed):

$$R_w = 0.43\ \text{m}^2\text{°C/W}$$

Vertical film coefficients:

$$h_1 = 8.3\ \text{W/m}^2\text{°C}$$

$$h_5 = 9.0\ \text{W/m}^2\text{°C}$$

$$h_6 = 9.3\ \text{W/m}^2\text{°C}$$

(cold ceiling, hot floor).

$$\text{ach} = 0.5$$

Calculation of infiltration conductance:

$$c_{pair} \equiv 1000\ \text{J/kg°C}$$

$$\rho_{air} \equiv 1.2\ \text{kg/m}^3$$

$$U_{inf} = [(\text{ach} \times \text{Vol})/3600\ \text{s}] \times \rho_{air} \times c_{pair} = 60\ \text{W/°C}$$

We then calculate all resistances as in the previous section.

Outside temperature

The outside temperature for a day is modelled by a Fourier series based on $N_{\mathrm{To}} + 1$ values that are an input to the array. If more detail is required, N_{To} may be increased. Then, the Fourier series may be used to generate intermediate values as required by the time step of a finite difference model.

Solar radiation (approximate model – for detailed model see Chapter 3)

Solar radiation transmitted through the windows may be modelled by a half-sinusoid from sunrise to sunset time t_s. For initial analysis of radiant heating, the solar radiation may be set to a low value (50 W/m² of window).

$$S_{\max} = 50 \text{ W/m}^2$$

$$S_{it} = S_{\max}\cos[p(t_{it} - 12 \text{ h})/2t_s]^+ \text{ (only positive part)}$$

S_{it} may be modelled with a discrete Fourier series as follows: we multiply by total window area to determine total solar radiation.

$$S_{n_n} = \left(\sum_{it} S_{it} \frac{\exp(-jw_n t_{it})}{24}\right)\sum_{iw} A_{w_{iw}}$$

FINITE DIFFERENCE MODEL (ATHIENITIS 1994)

The general form of the explicit finite difference formulation corresponding to node i and time interval p is:

$$T(i,p+1) = \left(\frac{\Delta t}{C_i}\right)\left(q_i + \sum_j \frac{T(j,p) - T(i,p)}{R(i,j)}\right) + T(i,p) \tag{7.9}$$

where C is capacitance, j represents all nodes connected to node i, and q is a heat source such as auxiliary heating or solar radiation.

Critical time step (for all nodes i):

$$\Delta t_{\text{critical}} = \min\left(\frac{C_i}{\sum_j \frac{1}{R_{i,j}}}\right) \tag{7.10}$$

(the selected time step should be lower to ensure numerical stability).

The thermal network is shown below. The floor is discretized into two layers (one thermal capacitance and two resistances for each layer). The unheated surfaces are represented by node 6. S represents solar radiation transmitted into the room and absorbed by the surfaces (a low value is assumed). Resistance R_{10} represents heat loss by infiltration and through the windows and doors.

Thermal capacitances and resistances:

$$C_{\text{floor}} = c_6 \rho_6 A_6 L_6$$

$$C_4 = C_{\text{floor}}/2$$

$$C_3 = C_4$$

Thermal capacitance of the interior layer of unheated surfaces:

$$C_5 = \Sigma_{\text{se}} A_{\text{se}} L_{\text{se}} c_{\text{se}} \rho_{\text{se}}$$

$$R_{\text{o}} = (1/u_6 A_6) + (L_6/4k_6 A_6)$$

$$R_{34} = L_6/2k_6 A_6$$

$$R_{23} = R_{34}/2$$

$$R_{12}\ 1/A_6 h_6$$

Radiation heat transfer coefficient between floor and unheated surfaces:

$$h_{6\text{r}} = 4\ \text{W/m}^2{}^\circ\text{C}$$

$$R_{56} = \frac{1}{\sum_{\text{se}} \frac{k_{\text{se}} A_{\text{se}} 2}{L_{\text{se}}}}$$ This represents the thermal resistance of half of the innermost layer of the unheated surfaces

$$R_{26} = \frac{1}{h_{6\text{r}} A_6}$$

$$R_{16} = \frac{1}{\sum_{\text{se}} A_{\text{se}} h_{\text{se}}}$$

$$R_{50} = R_{56} + \frac{1}{\sum_{\text{se}} \frac{A_{\text{se}}}{R_{\text{se}}}}$$

$$R_{10} = \frac{1}{U_{\text{inf}} + \sum_{ii} \left(\frac{A_{\text{w}_{ii}}}{R_{\text{w}}} + \frac{A_{\text{d}_{ii}}}{R_{\text{d}}} \right)}$$

STABILITY TEST TO SELECT TIME STEP

The time step Δt should be lower than the minimum of the three values in the vector TS:

$$\mathrm{TS} = \frac{C_3}{\left(\frac{1}{R_{23}} + \frac{1}{R_{34}}\right)} \quad \frac{C_4}{\frac{1}{R_0} + \frac{1}{R_{34}}} \quad \frac{C_5}{\left(\frac{1}{R_{56}} + \frac{1}{R_{50}}\right)}$$

$$\Delta t_{\text{critical}} = \min(\mathrm{TS}) = 215.686 \text{ s}$$

Select smaller time step than critical: $\Delta t = 200$ s.

The simulation will be performed for two days (periodic). Thus, weather data have to be generated for NT times as follows:

$$\mathrm{NT} = 86400 \text{ s}/\Delta t \times 2 = 864$$

(number of time steps for two days).

$$p = 0,1..\mathrm{NT}$$

Times at which simulation is to be performed:

$$t_p = p\Delta t$$

Assume basement temperature, $T_{bp} = 16°\mathrm{C}$.

Initial conditions must first be assumed. Also, the size of the heating system q_{max} has to be decided together with the proportional control constant K_p. A good estimate for q_{aux} can be determined by multiplying the zone conductance Y_{z0} with the maximum indoor–outdoor temperature differential and increasing the result by

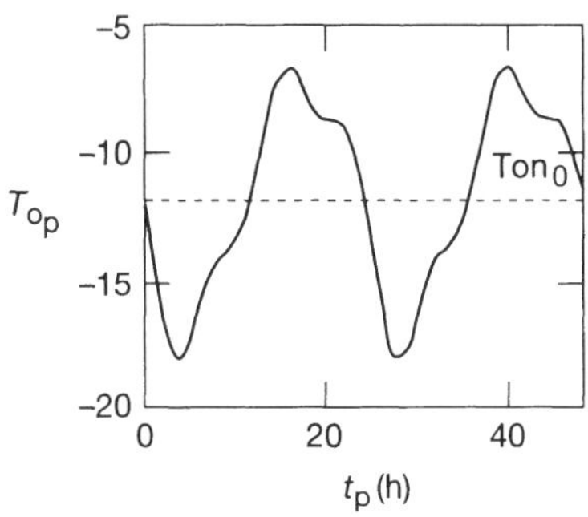

Figure 7.9 Ambient temperature (typical cold day profile (2% probability) for Montreal)

Figures 7.9 and 7.10 show the ambient temperature and solar radiation profiles employed in simulations.

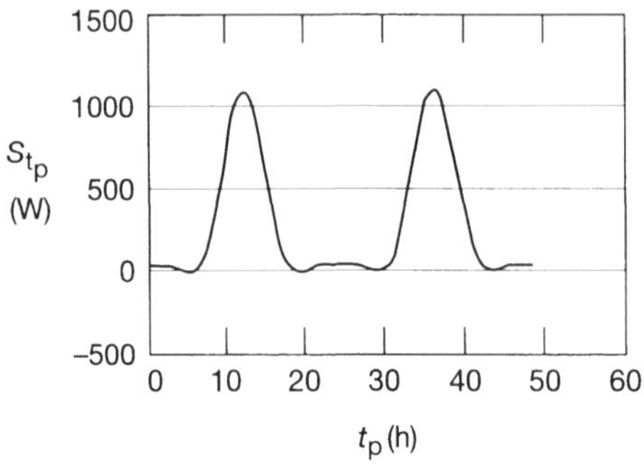

Figure 7.10 Solar radiation (total in zone)

50%. The transient room temperature response will then show whether this assumption was correct. For example:

$$q_{max} = |Y_{z0}| 30°C \times 1.5 = 1.054 \times 10^4 \text{ W}$$

Initial estimates of temperatures:

$$\begin{pmatrix} T_{1,0} \\ T_{2,0} \\ T_{3,0} \\ T_{4,0} \\ T_{5,0} \\ T_{6,0} \end{pmatrix} = \begin{pmatrix} 21 \\ 24 \\ 26 \\ 29 \\ 17 \\ 18 \end{pmatrix} °C$$

$$q_{aux} = 0 \text{ W}$$

Setpoint can vary with time of day if desired:

$$T_{sp} = 22°C$$

Proportional control constant (a good value is $q_{max}/2$):

$$K_p = 5000 \text{ W/°C}$$

The equations below were employed in Mathcad (Athienitis 1998) to determine the response of the system. The first equation describes the heating source (no heating is provided if the difference $(T_{sp} - T_1) < 0°C$).

$$\begin{pmatrix} q_{\text{aux}_{p+1}} \\ T_{1,p+1} \\ T_{2,p+1} \\ T_{3,p+1} \\ T_{4,p+1} \\ T_{5,p+1} \\ T_{6,p+1} \end{pmatrix} = \begin{bmatrix} \left(\text{if}\left[K_p\left(T_{\text{sp}} - T_{1,p}\right) > q_{\max}, q_{\max}, K_{\text{p}}\left(T_{\text{sp}} - T_{1,p}\right)\right]\left[\left(T_{\text{sp}} - T_{1,p}\right) > 1.0\right]\right) \\ \dfrac{\dfrac{T_{2,p}}{R_{12}} + \dfrac{To_p}{R_{10}} + \dfrac{T_{6,p}}{R_{16}} + 0.7St_p}{\left(\dfrac{1}{R_{12}} + \dfrac{1}{R_{10}} + \dfrac{1}{R_{16}}\right)} \\ \dfrac{\dfrac{T_{3,p}}{R_{23}} + \dfrac{T_{6,p}}{R_{26}} + \dfrac{T_{1,p}}{R_{12}}}{\left(\dfrac{1}{R_{23}} + \dfrac{1}{R_{26}} + \dfrac{1}{R_{12}}\right)} \\ \dfrac{\Delta t}{C_3}\left(\dfrac{T_{4,p} - T_{3,p}}{R_{34}} + \dfrac{T_{2,p} - T_{3,p}}{R_{23}}\right) + T_{3,p} \\ \dfrac{\Delta t}{C_4}\left(\dfrac{T_{b_p} - T_{4,p}}{R_0} + \dfrac{T_{3,p} - T_{4,p}}{R_{34}}\right) + T_{4,p} \\ \dfrac{\Delta t}{C_5}\left(\dfrac{T_{6,p} - T_{5,p}}{R_{56}} + \dfrac{T_{0_p} - T_{5,p}}{R_{50}}\right) + T_{5,p} \\ \dfrac{\dfrac{T_{2,p}}{R_{26}} + \dfrac{T_{5,p}}{R_{56}} + \dfrac{T_{1,p}}{R_{16}} + 0.3St_p}{\left(\dfrac{1}{R_{56}} + \dfrac{1}{R_{26}} + \dfrac{1}{R_{16}}\right)} \end{bmatrix}$$

The results for second day (first day is affected by assumed initial conditions) are given in Figures 7.11 and 7.12.

Heating energy consumption by numerical integration for second day:

$$Qh = \sum_{v} \frac{q_{\text{aux}_v} + \left|q_{\text{aux}_v}\right| + q_{\text{auxt}_{v+1}} + \left|q_{\text{aux}_{v+1}}\right|}{4}\left(t_{v+1} - t_v\right)$$

$$= 5.785 \times 10^8 \text{ J}$$

Peak heating load = 8.7 kW

The above load may be employed as a basis for sizing the system based on the extreme weather conditions assumed. Strategies for reduction of peak load include increasing the mass and preheating to a higher setpoint during the daytime when loads are lower than night-time on very cold days.

However, on sunny days this will reduce the amount of solar radiation that can be stored in the mass; on these days it is preferable to use a lower setpoint during the day and allow the floor mass to be heated by solar radiation transmitted through a south-facing window.

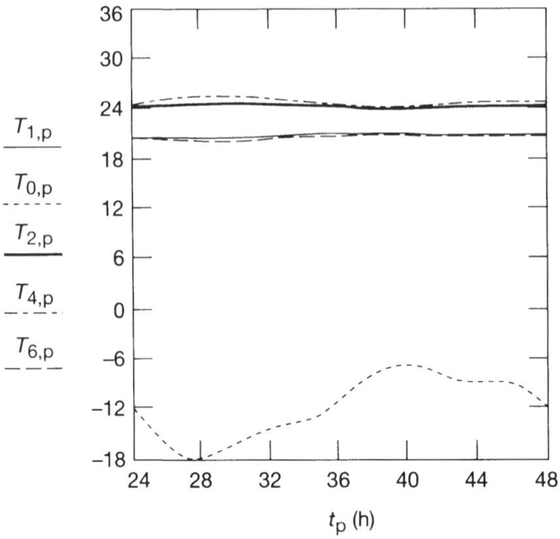

Figure 7.11 Temperatures (T_1 is room air temperature)

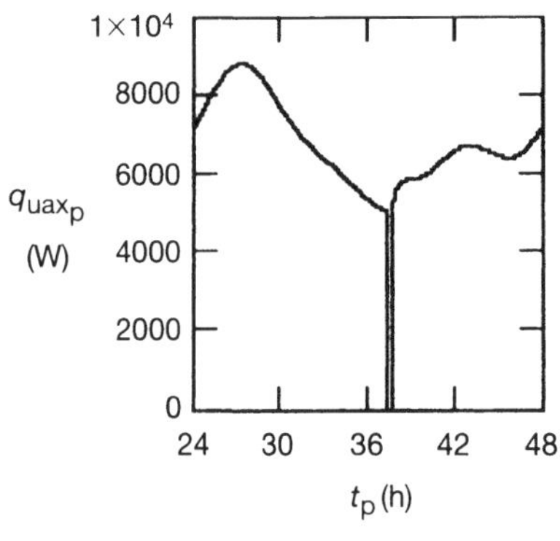

Figure 7.12 Heating profile

7.3 Analysis of integrated passive–active systems: a ground cooling/heating application

In most locations the temperature of the soil at depths of 2 m or more varies with small amplitude throughout the year. This provides an opportunity for passing air through long underground ducts to condition it in winter by warming it up and cooling it down in summer. This process can be particularly effective when the air is relatively dry.

Several researchers have studied the use of the ground for heat dissipation, for example Mihalakakou *et al.* (1994), who developed a transient numerical model of coupled heat and mass transfer for buried pipes and validated it with experimental data. This section considers combination of ground cooling/heating of air with a variable air volume (VAV) system. The basic idea is to preheat or precool the fresh air intake of the VAV system in order to reduce energy costs.

This study describes application of this principle to the design of three buildings (nos 11, 12, 13) of the new University of Cyprus campus, located at latitude 35°N with a mild temperate dry climate. The winter and summer design temperatures are typically about 0°C and 40°C respectively. However, the temperature of the soil at a depth of 2 m varies from a maximum of approximately 22°C in September to a minimum of 16°C at the end of February. Thus, given the dry climate, this provides an opportunity to cool room air in summer and heat it in winter by passing through underground channels.

7.3.1 Mathematical model for ground heat transfer

The basic objective of the model developed for the underground channels was to determine optimum dimensions so as to achieve maximum heat transfer between the air and the ground. At the same time, pressure drops needed to be reduced to a reasonable value so as not to require very large fans and waste much of the thermal energy potentially saved in fan energy requirements.

Fresh air will be drawn through wind towers (which will function as simple fresh air intakes); after passing it through the underground channels (at a depth of about 2 m) it will be fed into the fresh air intake of a variable air volume HVAC system. In designing the air channels under the buildings, the objective is to achieve maximum heat transfer and minimum pressure drop, while limiting the maximum velocity of air in the vertical ducts leading to the air-handling units to about 10 m/s. The design concept for the systems is shown in Figure 7.13.

Since the university will typically not operate late at night, fresh air requirements at that time will be minimal and the ground will be expected to recover its normal average temperature during the night. Thus, it is reasonable to assume approximately constant ground temperature for periods of operation of 10–15 hours, as also indicated by the results of Mihalakakou *et al.* (1994) (the temperature of the ground rises by only about 2°C after continuous operation in summer).

By performing an energy balance on a strip of width dx of the soil all around the underground channel, and air flowing over it, assuming that the soil (at depth D) in contact with the channel is at a constant surface temperature, the following ordinary differential equation is obtained:

$$V H W c \rho \, \mathrm{d}T = 2 f (H + W) \, \mathrm{d}x \, h (T_{\mathrm{w}} - T) \qquad (7.11)$$

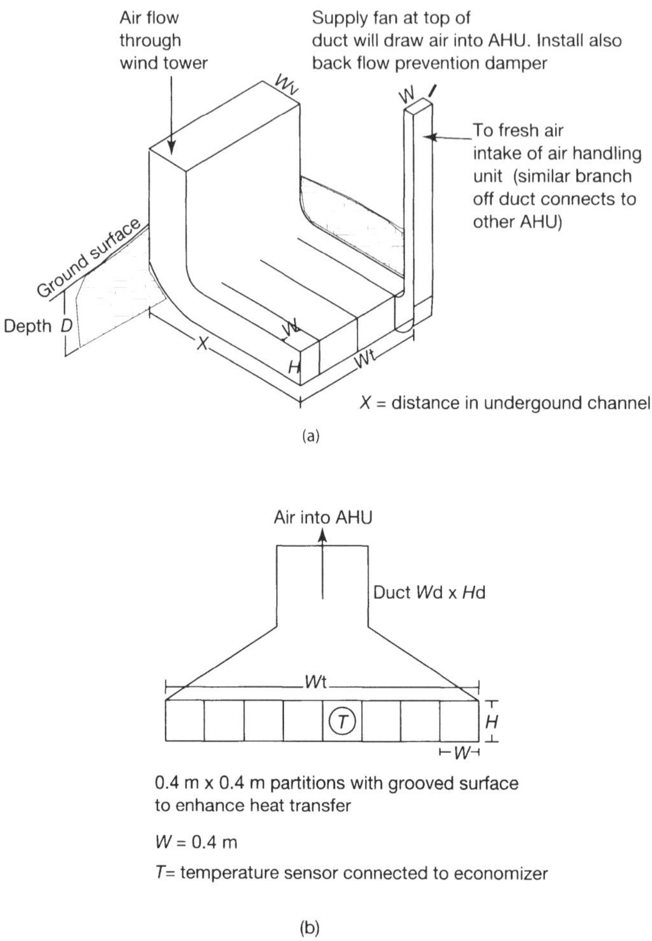

Figure 7.13 (a) Schematic of wind tower and underground tunnel concept for buildings (only one duct connecting to a typical air-handling unit (AHU) is shown); (b) section showing one underground channel connected to only one AHU (building 11)

where

V = air velocity

H = channel height

W = channel width

f = area increase factor if a corrugated surface is used

h = heat transfer coefficient

T_w = channel soil surface temperature (assumed constant)

T = air temperature as a function of distance travelled x

Thus, the differential equation can be written as

$$a\, \mathrm{d}(T - T_w)/\mathrm{d}x + (T - T_w) = 0 \tag{7.12}$$

where

$$a = VHWc\rho/[2f(H + W)h]$$

The solution for the above ordinary differential equation gives an exponential relationship for the air temperature:

$$T(x) = T_w + (T_o - T_w)\mathrm{e}^{-x/a} \tag{7.13}$$

The heat transfer coefficient h was determined using the Dittus–Boelter correlation (Holman 1986):

$$Nu = 0.023Re^{0.8}Pr^{n} \tag{7.14}$$

where Nu is Nusselt number, Re is Reynolds number and Pr the Prandtl number for air (Holman 1986); $n = 0.4$ for heating and 0.3 for cooling.

The temperature of the air rises as it flows through the tower in winter and drops in summer following an exponential curve. If the channel is long enough, the air temperature will approach the ground temperature, especially if the heat transfer area is high. However, if the flow diameters (hydraulic) are too small and the duct too long, the friction losses and pressure drops will be too high, requiring too large a supply fan to draw the air into the air handling unit.

Figure 7.14 shows typical air temperature rise with distance travelled in winter and Figure 7.15 the air temperature drop in summer. Figure 7.16 depicts typical pressure losses for the air as it flows through the underground channel. Correlations suggested by ASHRAE (1997) were utilized for the pressure drop calculations. These correlations may be employed for pressure drop in air duct systems.

The optimum air speed range when the size of the AHUs, pressure drops and ground heat transfer were considered was found to be 2–3 m/s.

For building 11, which will consist primarily of classrooms with high fresh air requirements (served by one AHU), the optimum height (H) of the channel (see Figure 7.13) was 0.4 m; the width (W_t) was fixed at 3.6 m and the length was set at about 40 m based on architectural constraints. The AHU was at the opposite end from the wind tower, thus creating an optimum arrangement for heat transfer from the soil. The channel will be subdivided into eight sections to enhance heat transfer (Figure 7.13b). The vertical duct connecting to the AHU will have a section of about 1.00 m × 0.60 m, delivering fresh air at a maximum speed of about 9 m/s.

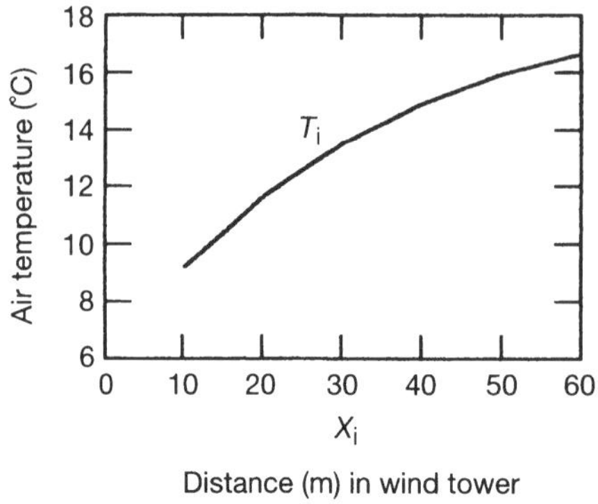

Figure 7.14 Temperature rise of the air as it flows through the underground channel in winter (outside temperature = 6°C, ground temperature = 19°C, air speed 3 m/s)

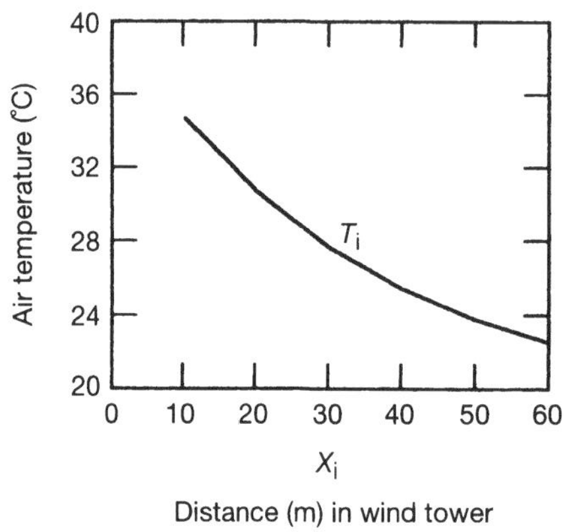

Figure 7.15 Temperature drop of the air as it flows through the channel in summer (outside temperature = 40°C, ground temperature = 19°C, air speed 3 m/s)

In the case of buildings 12 and 13 the length of the underground channel is 60–70 m but several AHUs were to be situated in the centre of the buildings. Thus, with air intake from both sides of the channel, the path travelled underground will be 30–40 m. As can be seen from Figure 7.15, with a distance travelled of 40 m we have a substantial cooling of air from 40°C to about 25°C (ground temperature 19°C). Thus, this length is adequate (for building 11) to achieve a temperature that is close to what would be obtained with a channel of infinite length (limiting case, in which the air comes out at ground temperature).

Figure 7.17 shows the temperature and pressure profiles for building 11.

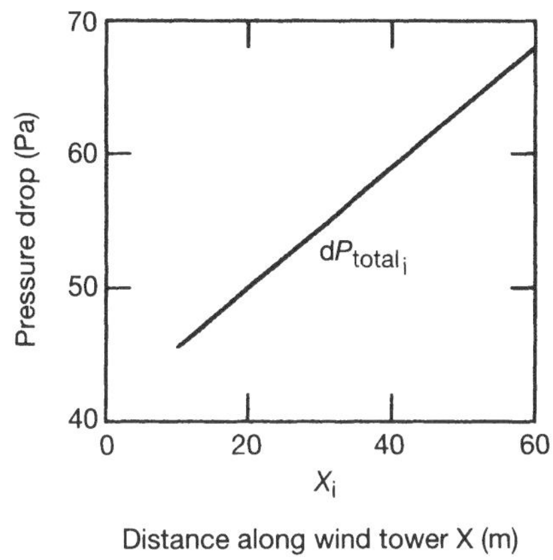

Figure 7.16 Pressure losses (Pa) of the air as it flows through the channel in summer (air speed 3 m/s)

7.3.2 Simulation results

Detailed simulation results are described for building 11. This building has a floor area of about 1200 m^2. It consists primarily of classrooms and generally has a lower fraction of window area than buildings 12 and 13. Thus, the solar gains are lower but it has high internal gains owing to the large number of occupants (maximum about 800) and high fresh air requirements (minimum 6000 l/s). The internal gains were assumed to be about 63 kW between 8 a.m. and 5 p.m.

Peak refrigeration loads and heating loads were calculated by selecting extreme design days in July and January respectively. For the refrigeration load calculation, a design day with maximum temperature of 40.5°C and minimum 25.5°C was chosen. For the heating calculations, the outdoor temperature varied from a minimum of 0°C to a maximum of 10°C. The peak load is dictated primarily by the maximum temperature for cooling and the minimum temperature for heating. In both cases a sinusoidal temperature profile was employed. For cooling, high solar gains and internal gains were used, while for heating medium gains were assumed. The calculation was dynamic, taking into account heat storage in the building envelope. A dynamic thermal analysis simulation program based on the model described in Section 7.1 was employed (Athienitis 1994b) modified to represent ventilation and solar radiation in detail. A supply air temperature of about 13°C was assumed for cooling and 24–27°C for heating. Indoor temperature was set to 24°C for cooling and 20°C for heating (18°C at night).

Detailed dynamic thermal simulations were performed for an average day of each month of the academic year to evaluate the potential energy savings due to the underground channels as well as the thermal mass of the building. The thermal mass

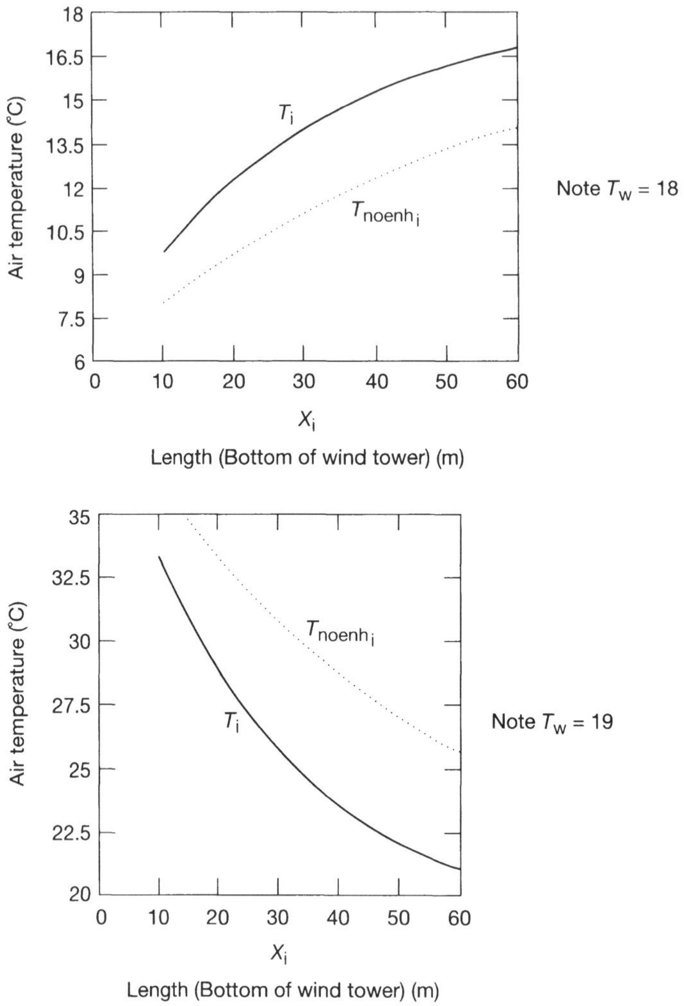

Figure 7.17 Windtower-channel temperature profiles for air preheating and air precooling

generally reduces peak loads. Table 7.1 presents summary results for building 11.

The peak refrigeration load (see Figure 7.18) is about 142 kW. The maximum quantity of supplied air required is about 10–12 m^3/s. Without fresh air precooled by the underground channels, the peak load rises by about 90–100 kW. When internal reflective rolling shutters are used the peak load is reduced by 10 kW. The energy consumption is reduced by about 500 MJ (per day).

In Figure 7.18 the air circulation rate at night is set to a minimum of two air changes per hour. During the period 8 a.m. to 5 p.m. the minimum circulation rate,

Table 7.1 Energy analysis for building 11 (for average day in each month; note: thermal energy consumption only, negative denotes cooling, soil temperature is at 2 m depth)

Month	*Mean outdoor temp. (°C)*	*Soil temp. (°C)*	*Energy (MJ) With ground channel*	*Energy (MJ) Without ground channel*	*Typical peak load (kW)*
September	25.9	23	–4602	–5975	–124
October	20.8	22.5	–2829	–3310	–92
November	15.9	21	635	1004	47
December	11.9	19	1969	2683	82
January	10.1	16	3155	3550	105
February	10.4	15.5	2884	3367	94
March	12.4	16	1990	2205	77
April	17.3	16.5	–1144	–1934	–36
May	22.0	18	–2460	–4000	–73
June	25.6	19	–3523	–5603	–98

Total heating energy with channel = 10630 MJ
Total heating energy without channel = 12810 MJ
Total cooling energy with channel = 14560 MJ
Total cooling energy without channel = 20820 MJ
Assumptions in energy simulations:

- Internal gains are 70% of the maximum on an average day during the cooling season.
- Internal gains are 50% of maximum on an average day during the heating season.
- Reflective roller shades are not used very much i.e. solar gains are about 60–70% of those on a clear day.
- Thermostat setpoint for heating 20°C (16°C at night).
- Thermostat setpoint for cooling 24°C.
- Fresh air is obtained either from the wind tower or from outside by comparing the two temperatures in the economizer of the HVAC system. For example, in October it is often preferable to use outdoor air directly.

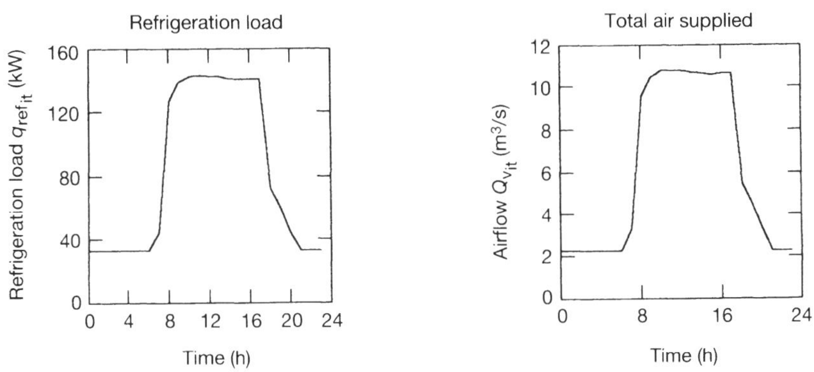

Figure 7.18 Refrigeration load (maximum) and supplied air for building 11. Outside temperature: maximum = 40.5°C, minimum = 25.5°C.

which is 100% fresh air from the wind tower, is about 6000 l/s. The fresh air is normally all supplied from the wind tower precooled to 24–27°C, i.e. close to the temperature of the return air from the zone, with which it is mixed in the AHU. Thus, if the temperature of the air supplied from the wind tower is less than 24°C then it may be beneficial to draw even higher amounts.

As can be seen from Table 7.1, the cooling energy consumption is about 60% higher than the heating energy consumption (without counting July and August). The energy consumption without the channel is about 40% higher for cooling and 20% higher for heating.

The highest refrigeration load with ground cooling is in September and not in June. The reason for this difference is the lower soil temperature in June, which increases the effectiveness of the ground channels by precooling the fresh air to a lower temperature than in September. Similarly, the highest heating savings are observed for December when the soil temperature is still high. The temperature of the soil at a depth of 2 m varies a few degrees about an annual average of 19°C. Because of the large thermal lag, the highest soil temperature is in September.

In calculating the energy consumption in Table 7.1, the temperature of the outside air is compared with that which would be obtained from the underground channel and the preferred source of fresh air is chosen. This strategy will also be implemented in the economizer controls of the variable air volume systems.

Generally the air-conditioning system sizes are dictated by cooling loads for this climate. However, heating loads are also important, particularly in the early mornings in January and February. The peak heating loads were determined as follows for building 11: a peak heating load of 125 kW in January and an energy consumption of 5720 MJ (see Figure 7.19). Without preheated air from the wind tower, the peak load is 171 kW and the energy consumption 7223 MJ.

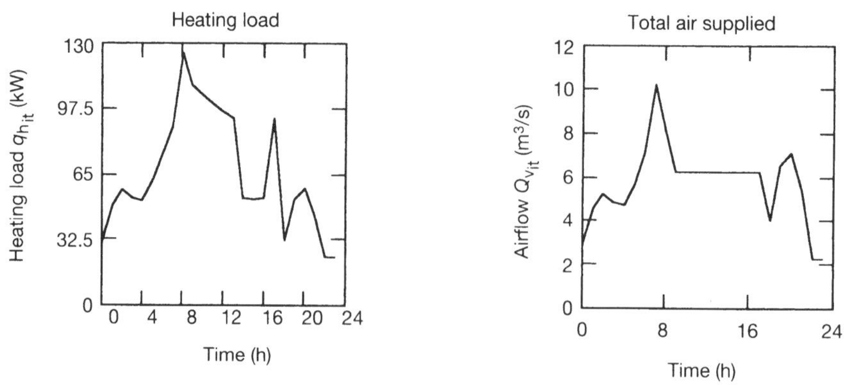

Figure 7.19 Heating load (maximum) and supplied air for building 11. Outside temperature: maximum = 10°C, minimum = 0°C

In all simulations a conservative approach was employed in evaluating the potential benefits due to the wind tower; for heating it was assumed that the air from the towers reaches a temperature equal to the average of the outdoor and deep ground temperatures (see Table 7.1). For cooling it was assumed to reach a temperature equal to $2/3 \times T_{soil} + 1/3 \times T_{exterior}$. As can be seen from Figures 7.14 and 7.15, these assumptions are justified. Note that the peak of the heating curves occurs in the early morning and is of short duration. The flat part of the air supply curves indicates the minimum amount of air required to satisfy the fresh air requirements.

The dynamic variation of the load and airflow for average days in June and January are shown in Figures 7.20 and 7.21 respectively. Comparison of Figure 7.18 (peak summer day) with Figure 7.20 (average summer day) indicates a substantially lower load on an average summer day, and hence less demand for

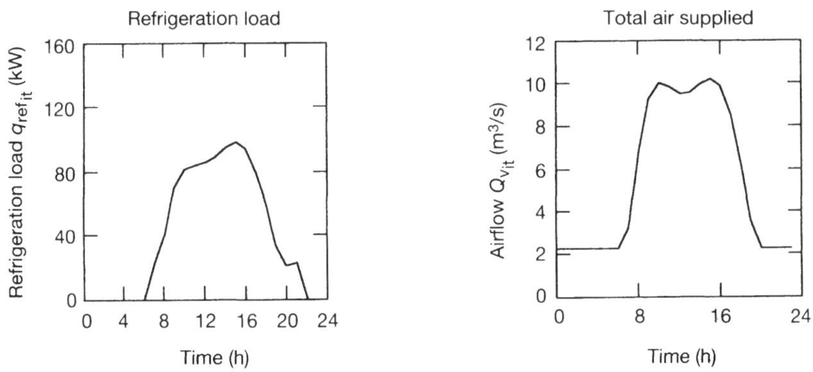

Figure 7.20 Refrigeration load and airflow for an average day in June for building 11 (at night there are a minimum of two air changes per hour)

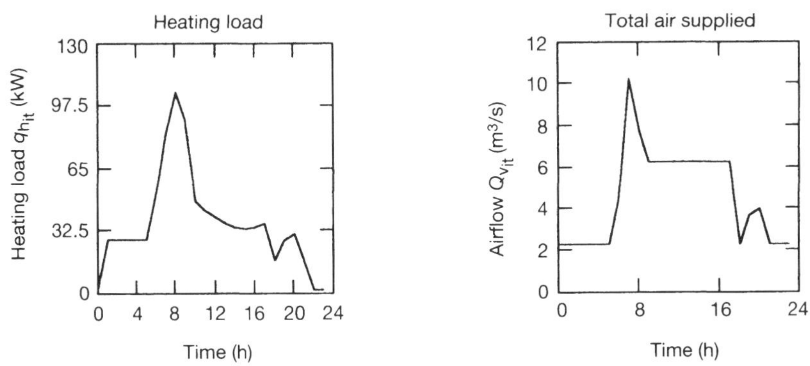

Figure 7.21 Heating load and airflow for an average day in January for building 11 (at night we have minimum two air changes per hour)

ground cooling. An analogous observation is made by comparing Figures 7.19 (peak winter day) and 7.21 (average winter day); the peak heating demand has a very short duration (a couple of hours) in the early morning, and the ground preheating of the fresh air will be very effective for this short period.

7.4 Bibliography

ASHRAE (1997) *ASHRAE Handbook 1997 Fundamentals*, American Society of Heating, Refrigerating, and Air-Conditioning Engineers, Inc., Atlanta, GA.

Athienitis A.K. (1994a) *Building Thermal Analysis*, Electronic Mathcad Book, MathSoft Inc., Boston, MA.

Athienitis A.K. (1994b) 'Numerical model for a floor heating system', *ASHRAE Transactions*, **100**:1, 1024–1030.

Athienitis A.K. (1997) 'Theoretical investigation of thermal performance of a passive solar building with floor radiant heating', *Solar Energy*, **61**:5, 337–345.

Athienitis A.K. (1998) *Building Thermal Analysis*, Electronic Mathcad Book, 2nd edn, Mathsoft Inc., Boston, USA.

Athienitis A.K., Stylianou M. and Shou J. (1990) 'A methodology for building thermal dynamics studies and control applications', *ASHRAE Transactions*, **96**, Part 2, 839–848.

Holman J.P. (1986) *Heat Transfer*, McGraw-Hill Inc., New York.

Mihalakakou G., Santamouris M. and Asimakopoulos D. (1994) 'Modelling the thermal performance of earth-to-air heat exchangers', *Solar Energy*, **53**:3, 301–305.

CHAPTER 8

Control of passive solar buildings

Before considering the dynamic thermal control of passive solar buildings, it is necessary to establish the desired thermal comfort parameters and their acceptable ranges. Thermal comfort is discussed in Section 8.1, while a detail dynamic model for control is discussed in Section 8.2.

8.1 Thermal comfort

Comfort is affected by several factors: physical health, mental health, social factors, environmental factors, etc. Considering only the environmental factors, one must analyse the thermal environment, the visual factors, the acoustical factors and indoor air quality to be able to assess human comfort. Thermal comfort is basically a condition of mind that expresses satisfaction with the thermal environment. In order to achieve good thermal comfort, many personal and environmental conditions must be fulfilled, ranging from physical to physiological and psychological (see Figure 8.1).

The main factor affecting thermal comfort is the temperature. However, the temperature is not the only environmental factor affecting thermal comfort. The thermal environment is a combination of air temperature, radiant temperature, air velocity and air humidity. These thermal factors and two personal factors (clothing insulation and activity level) are the main parameters influencing thermal comfort. In a solar building, we may also have diffuse and possibly beam solar radiation incident on a person.

The human body produces heat by metabolism (i.e. oxidation of food), and exchanges heat with the environment through the skin and by respiration. Under neutral conditions, for sedentary activities the heat exchange results in an internal body temperature of about 37°C, a skin temperature between 33°C and 34°C, and negligible skin moisture content (ASHRAE 1997). If there is any thermal disturbance to this neutral condition, the body's temperature control system makes an effort to maintain the neutral temperature (37°C) for the vital organs. The body heat balance at any moment results in an internal body temperature, a skin temperature and skin moisture content that are felt (sensed) by the brain. The mind

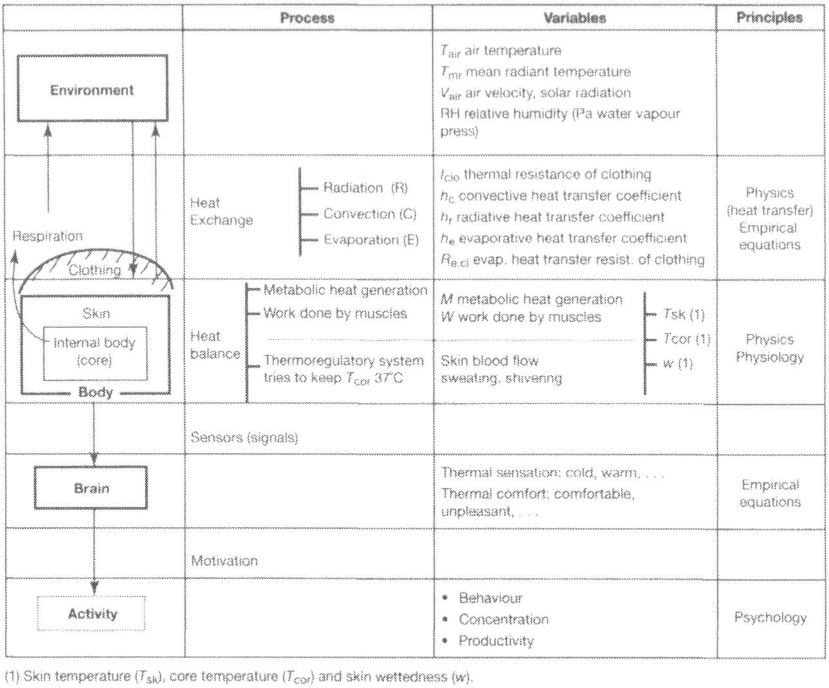

Figure 8.1 Schematic showing mechanisms affecting thermal comfort

perceives the thermal sensation (cold, warm, etc.) and draws conclusions about the comfort level (pleasant, unpleasant, etc.).

Simultaneously, the body's temperature control system (thermoregulatory system) tries to maintain the normal temperature of vital organs by different means depending on the magnitude of the thermal disturbance. When low-magnitude disturbances occur, the vasomotor control system supplies more blood to the extremities to lose heat (when warm), or less blood to maintain the heat in the vital organs (when cold). If the magnitude of the disturbance increases, the thermal sensation forces a behavioural response from the body (e.g. take off sweater, move to the shade when warm). If the disturbance increases even more, the body tries to lose more heat by evaporative sweating (when warm) or increase the heat generation by shivering (when cold).

Metabolic heat generation varies directly with the activity level: higher activity levels generate more heat, and lower activity levels less heat. Researchers (Gagge *et al.* 1973) have found that, for sedentary activities, under cold conditions the mean skin temperature is a good indicator of cold discomfort and usually ranges between 28°C and 33°C. Under warm conditions the mean skin temperature becomes a poor indicator of warm discomfort; it varies between 34°C and 36°C (only two degrees).

The 2 deg C difference limit of change in mean skin temperature is caused by evaporation from sweating and, therefore, the skin wetness becomes the best indicator of warm discomfort.

8.1.1 Body heat balance

The metabolic energy (M) produced by the body is expended in part as work (W) done by the muscles, and the difference is stored (S) as heat in the body. The stored heat (S) is dissipated to the environment through the skin surface and the respiratory tract. Because the heat exchange between the body and the environment occurs mostly through the skin, all heat balance variables are expressed per unit area of skin. Therefore, the heat balance equation is as follows.

$$M - W = Q_{sk} + Q_{res} + S \tag{8.1}$$

$$= (C + R + E_{sk}) + (C_{res} + E_{res}) + S \tag{8.2}$$

where
M = rate of metabolic heat rate production, W/m^2
W = rate of mechanical work accomplished, W/m^2
Q_{sk} = total rate of heat loss from skin, W/m^2
Q_{res} = total rate of heat loss through respiration, W/m^2
$C + R$ = sensible heat loss from skin by convection and radiation
E_{sk} = rate of total evaporative heat loss from skin, W/m^2
C_{res} = rate of convective heat loss from respiration, W/m^2
E_{res} = rate of evaporative heat loss from respiration, W/m^2
S = rate of heat storage in the body, W/m^2

8.1.2 Thermal exchange with the environment

Thermal exchange with the environment occurs in the form of sensible and latent heat through the skin and by respiration.

THERMAL EXCHANGE THROUGH THE SKIN

This takes the form of sensible heat by convection (C) and radiation (R), latent heat flow from evaporation of sweat (E_{rsw}), and from evaporation of moisture diffused through the skin (E_{dif}). For the sensible heat exchange by convection and radiation the heat exchange can be expressed in terms of the convective and radiative heat transfer coefficients, as follows:

$$\begin{aligned} C &= f_{cl} h_c (T_{cl} - T_a) \\ R &= f_{cl} h_r (T_{cl} - T_{mr}) \end{aligned} \tag{8.3}$$

where:
h_c = convective heat transfer coefficient, W/m^2 K
h_r = radiative heat transfer coefficient, W/m^2 K
f_{cl} = body area correction factor area clothed body/area nude body (A_{cl}/A_D)
T_{cl} = mean temperature of the outer surface of the clothed body
T_a = air temperature
T_{mr} = mean radiant temperature from the surrounding surfaces

The total sensible heat exchange is obtained by adding C and R as follows:

$$C + R = f_{cl}h(T_{cl} - T_o) \tag{8.4}$$

where:

$$T_e = \frac{h_r\, T_{mr} + h_c\, T_a}{h_r + h_c} \tag{8.5}$$

$$h = h_r + h_c \tag{8.6}$$

T_e is defined as the operative temperature. It is the average of the air temperature and the mean radiant temperature, weighted by their respective heat transfer coefficients, and it is the closest approximation to the actual temperature as felt by a person.

h is combined convective (h_c) and radiative (h_r) heat transfer coefficient, W/m^2 K.

The above equations relate the heat exchange to the clothing surface temperature. It is more convenient to relate them to the mean skin temperature. The sensible heat through the clothing can be written as a function of the thermal resistance of clothing:

$$C + R = (T_{sk} - T_{cl})/\mathrm{R}_{cl} \tag{8.7}$$

Therefore, combining equations 8.4 and 8.7 to eliminate T_{cl} the following steady-state sensible heat-exchange equation is obtained:

$$q\,/\,A_{sens} = h_{sens}\Delta T \quad (C+R) = \frac{1}{R_{cl} + \dfrac{1}{(f_{cl}\; h)}}\left(T_{sk} - T_e\right) \tag{8.8}$$

where
R_{cl} = thermal resistance of clothing, m^2 K/W
T_{sk} = mean skin temperature, °C

Similarly for the latent heat exchange the following equation is obtained:

$$Q/A_{latent} = h_{latent}\Delta p \tag{8.9}$$

$$E_{sk} = \frac{w}{R_{e,cl} + \frac{1}{(f_{cl}\, h_e)}}(P_{sk} - P_a) \qquad (8.10)$$

where:

$E_{sk} = E_{rsw} + E_{dif}$

E_{sk} = rate of total evaporative heat loss from skin, W/m^2

w = wet skin fraction, dimensionless (skin wettedness)

$R_{e,cl}$ = evaporative heat transfer resistance of clothing (analogous to R_{cl}), m^2 kPa/W

h_e = evaporative heat transfer coefficient (analogous to h_c), W/m^2 kPa

P_{sk} = water vapour pressure at skin, normally assumed to be that of saturated water vapour at T_{sk}, kPa

P_a = water vapour pressure in ambient air, kPa

E_{rsw} = rate of evaporative heat loss of regulatory sweat, W/m^2

E_{dif} = rate of evaporative heat loss of natural diffusion water through skin, W/m^2

Empirical expressions are used to determine the values of the heat and moisture heat transfer coefficients for different personal and environmental situations. According to ASHRAE (1997) 'h_r is nearly constant for typical indoor temperatures, and a value of 4.7 W/m^2 K suffices for most calculations'. This value is affected by the parameter f_D = 0.73 for a standing person (Dubois effective body area factor (ASHRAE 1997)).

THERMAL EXCHANGE BY RESPIRATION

Thermal exchange by respiration is the sensible heat flow during respiration (C_{res}) and latent heat flow due to evaporation of moisture during respiration (E_{res}). Respiratory heat losses are expressed in terms of pulmonary ventilation rate, humidity ratio of exhaled and inhaled air, specific heat of air and other variables. After some approximations the following equation is obtained (ASHRAE 1997):

$$C_{res} + E_{res} = 0.014M(34 - t_a) + 0.0175M(5.87 - P_a) \qquad (8.11)$$

The convective and evaporative heat losses are then given in terms of the metabolic heat production (W/m^2), the air temperature (T_a, °C), and the water vapour pressure in the ambient air (P_a, KPa).

8.1.3 Thermal comfort models

Among the various types of thermal comfort model two have been utilized most widely: Fanger's steady-state energy balance model, and Gagge's two-node transient model. These models apply the energy balance equation, and estimate thermal comfort by empirical equations, under different assumptions. The thermal comfort predicted by these two models is general thermal comfort. Therefore, a neutral thermal sensation means that a person feels thermally neutral for the body as a whole, not knowing whether they would prefer a higher or lower ambient temperature level.

Fanger's model utilizes the PMV index (predicted mean vote) as the parameter to evaluate thermal sensation. The PMV index predicts the mean response of a large group of people according to the following thermal sensation scale: ±3 (hot/cold), ±2 (warm/cool), ±1 (slightly warm/cool), 0 (neutral). Gagge's model predicts thermal sensation by TSENS index, and discomfort by DISC index. TSENS is based on the PMV scale, but with extra terms for ±4 (very hot/cold) and ±5 (intolerably hot/cold). DISC recognizes the same positive/negative convention for warm/cold discomfort with the following scale: ±5 intolerable, ±4 limited tolerance, ±3 very uncomfortable, ±2 uncomfortable and unpleasant, ±1 slightly uncomfortable but acceptable, 0 neutral.

For most purposes these two models are sufficient to predict thermal comfort. In fact ISO comfort standard 7730 (1984) is based on thermal comfort results calculated by the Fanger model; and ASHRAE standard 55 (1992) is based on the results obtained from Gagge's model. However, these models have some limitations. Fanger's steady-state model is applicable for mainly sedentary people and assumes that the body is in steady-state thermal equilibrium. Therefore, vasoregulation, shivering and sweating are not considered. The model represents the body as a point, therefore the same skin temperature is assumed for the whole body. Internal body temperature (core temperature) is not considered. Gagge's two-node model is applicable for mainly sedentary people and assumes the body is divided by two concentric cylinders: the internal one is the core and the external one is the skin or shell. The temperature in each cylinder is assumed to be uniform, so that the only temperature gradients occur between cylinders.

8.1.4 Thermal non-uniform conditions

The above thermal comfort models calculate thermal comfort for the body as a whole. However, the body may be thermally neutral but at the same time any part of it may be too cold or too warm. Therefore, it is a further requirement for thermal comfort that no local warm or cold discomfort exists at any part of the body. Such a local discomfort may be caused by:

- asymmetric radiation
- draught: undesired local cooling of the body caused by air movement
- vertical air temperature gradients
- hot or cold floors.

To account for these non-uniformities of the environment the standards (ASHRAE 55-92, ISO 7730) specify limits for them. The effect of solar radiation is not considered in these standards.

THERMAL COMFORT STANDARDS

ASHRAE standard 55 (1992) and ISO standard 7730 (1984) are the basis for the design and operation of buildings' HVAC systems. Both standards specify the thermal environmental conditions for thermal comfort in buildings.

An acceptable thermal environment is defined in both ASHRAE standard 55-92 and ISO standard 7730 as an environment that at least 80% of the occupants would find thermally acceptable. Therefore, it is assumed that less than 20% will be dissatisfied when the recommendations of the standard are followed. The 20% dissatisfaction criterion includes dissatisfaction due to local and general thermal discomfort. In ISO standard 7730 the recommended limits for an acceptable thermal environment are:

$$-0.5 < \text{PMV} < 0.5$$

PPD < 10% (percent dissatisfied)

The standard recommends **operative temperature** intervals for different combinations of clothing insulation and activity level. The recommended relative humidity ranges from 30% to 70%. ASHRAE standard 55-92 recommends also the required **operative temperature** and relative humidity combinations for primarily sedentary activity ((1 met = 58.2 W/m^2), during winter and summer seasons with typical indoor clothing for each season as shown in Table 8.1.

Table 8.1 ASHRAE standard 55. Optimum acceptable ranges of operative temperature for people during light, primarily sedentary activity (≤1.2 met) at 50% relative humidity and mean air speed (0.15 m/s)

Season	*Typical clothing I_{cl} (clo)*	*Optimum operative temp. (°C)*	*Operative temp. range (10% dissatisfaction criterion) (°C)*
Winter	0.9	22	20–23.5
Summer	0.5	24.5	23–26
	minimal 0.05	27	26–29

In Table 8.1 the difference between summer and winter recommended operative temperatures is due to the lighter indoor clothing worn during the summer. ASHRAE also recommends a maximum relative humidity of 60% and a minimum humidity given by a dew point temperature of 2°C.

8.1.5 Local thermal discomfort

In ISO standard 7730 and ASHRAE standard 55-92 the recommended limits for local discomfort for people occupied with light, mainly sedentary activity (1.2 met) are as follows:

- The radiant temperature asymmetry from windows or other cold vertical surfaces shall be less than 10°C.
- The radiant temperature asymmetry from a heated ceiling must be less than 5°C.
- Vertical air temperature difference between head and ankle level shall be less than 3 deg C. In ASHRAE standard 55-92 it is specified as the difference between the air temperature at the 0.1 m and 1.7 m levels.
- Surface temperature of the floor shall normally be between 19°C and 26°C, but floor heating systems can be designed for 29°C. In ASHRAE 55-92 a general range of 18–29°C is specified.
- Draught can be the most critical variable because it is more difficult to control. It depends on many factors such as: air distribution system, air turbulence, position of furniture, air temperature, position of the person within the space, area of skin exposed, etc.

Both standards present in charts the allowable mean air velocity as a function of air temperature, and turbulence intensity that result in less than 15% dissatisfaction. Both standards also present in a graphic way the required air velocities to offset elevated temperatures above 26°C. However, the maximum allowable mean air velocities specified by ISO 7730 and ASHRAE 55-92 are 1 m/s and 0.8 m/s respectively, for sedentary activities. According to ASHRAE 55-92 'higher air speeds may be acceptable if the person has individual control of the local air speed'.

Finally ASHRAE 55-92 states that 'more active people are less sensitive, and they will be able to accept higher degrees of non-uniformity than people with light, primarily sedentary activity'. Some laboratory-based studies (Knudsen *et al.* 1989) recommend an optimal operative temperature in winter for a person undergoing sedentary activities (1.2 met) and who has a clothing level of 1 clo (typical indoor winter ensemble) as 22 ± 2°C. During summer warm condition with light clothing (0.5 clo) they recommend an operative temperature equal to 24.5 ± 1.5°C. However, in a field-based research, Griffiths (1990) shows that, generally, the temperatures required for thermal comfort are significantly lower in all building types than those predicted from models established through laboratory-based researches. He gave, for instance, the following operative temperature for optimal conditions: 21°C for office workers in Germany and France; 19°C for schoolteachers in France; 19°C for hospital occupants in the UK; and still lower temperatures for houses in the UK and France.

8.1.6 Effect of solar radiation

Solar radiation transmitted through fenestration has two effects on the indoor thermal environment: (1) immediate effects due to the incidence of direct and diffuse transmitted solar radiation on people, and (2) indirect effects due to absorption of part of the solar radiation by the room interior surfaces and furnishings and subsequent re-radiation (long wave) or convection from these surfaces.

The amount of direct solar radiation absorbed by a person depends on the position of the sun, the window properties and the person's average solar absorptance. Gianccone and Gianfranco (1986) developed a computer program to calculate thermal comfort maps that show the influence of the internal distribution of hourly radiation on thermal comfort taking into account the presence of direct, diffuse and reflected radiation falling on the occupant. Solar radiation was modelled with two different methods: (1) as uniformly distributed on all room surfaces, and (2) its internal distribution was accurately calculated taking into account the hourly path of solar radiation and external as well as internal shadowing effects. The two methods showed great differences in their estimation of thermal comfort. The PMV model was applied to estimate the thermal comfort level (Fanger 1970).

The *CIBSE Guide* (1987) states that the only significant component of transmitted solar radiation (for thermal comfort calculations) is the direct component. It gives the effect of varying levels of direct solar radiation on effective mean radiant temperatures. For example, for a clothing/skin absorptance of 0.8 and incident solar radiation equal to 500 W/m^2, a rise of 14.4°C in the effective mean radiant temperature is reported (equivalent to 7.7°C in the dry resultant temperature – which is what would be approximately measured with a globe thermometer having the same absorptance). Similar results have been reported by Olesen *et al.* (1989).

A study by Athienitis and Haghighat (1992) used experiments and simulation to further quantify both the direct and indirect effects of solar radiation on thermal comfort. An approximate form of the equation recommended for the effect of solar radiation is to increase the effective operative temperature as follows:

$$T_e = [h_r T_{mrt} + h_c T_{ai} + \alpha_s (q_d + q_b/4)]/(h_r + h_c)$$

where α_s is the skin/clothing average solar absorbtance. Note that the beam radiation q_b is divided by 4 because the projected area of a sphere (approximating a person) is a quarter of its surface area; it is assumed that the diffuse radiation q_d is received uniformly from all sides (acceptable for a lightly coloured room). The approximate effect of uniform diffuse radiation is thus q_d/10°C and for beam q_b/40°C. For example, 400 W of beam radiation transmitted through a window and incident on a person would cause an apparent increase of approximately 10°C in the operative temperature.

In the case of daylighting, we must also consider glare discomfort and control shading and light distribution devices to reduce the non-uniformity in daylight distribution (see also Sections 3.4 and 3.5).

8.2 Control analysis of passive solar buildings

8.2.1 Introduction

This section presents an integrated methodology for dynamic thermal analysis of passive solar buildings and their auxiliary HVAC systems. The methodology is based on the use of Laplace transfer functions for the building, its heating–cooling system and control components. These transfer functions can be used for building thermal control studies, frequency domain analysis and energy analysis. Laplace transfer functions for the building are obtained by means of thermal network models that include both distributed parameter elements such as thermal mass and lumped elements such as the room lightweight contents. The radiant exchanges are represented accurately. The methodology is applied to obtain both air temperature and operative temperature transfer functions. Laplace transfer functions are also used for HVAC system and control components. Transient thermal control studies are performed by means of an efficient numerical Laplace transform inversion technique. Building heating/cooling load calculations may be performed by means of discrete Fourier series techniques.

Building thermal dynamics and control studies are usually performed separately from HVAC system design and building energy analysis without sufficient consideration of the system loads, their variability and the interactions between the building shell, the HVAC system and the control system. These interactions and load variability are even higher in today's buildings because of the trend towards larger window areas and other passive solar design features in both commercial and residential buildings, the implementation of novel heating/cooling systems coupled with energy conservation measures and increased ventilation for improved indoor air quality. The interaction between the three building thermal subsystems (envelope, HVAC system, control system) is a complex process, which leads to delays in the building response. The order of magnitude of the thermal time lags introduced by the building envelope is of the order of hours, and sometimes a day or more for passive solar buildings. In general, high radiant gains have delayed effects that influence both load and room temperature significantly as the amount of building interior thermal mass increases. Time delays and thermal lags are also introduced by the HVAC system and the sensor–control system. The order of magnitude of these time delays is typically minutes, since time constants of heating/cooling coils (Bhargava *et al.* 1975) are usually smaller than a minute (Shavit *et al.* 1982), transportation times within ducts and pipes usually take less than 10 seconds, and sensor time constants are rarely greater than 1 min (Omega Engineering 1987). The time lags introduced by the building, its HVAC system, and the sensor–control system are one of the major causes of complexity in controlling indoor environments.

Neglect of the interaction between a building and its HVAC system and of the combined effect of the various time delays in building design may lead to comfort problems during building operation. Building operation may be made more energy efficient and comfortable by fully utilizing the capabilities of the new digital controllers that permit continuous modification of operating parameters, such as setpoints, as a result of predicted and sensed performance. To achieve this, one must address the interaction between the long-term building dynamics associated with the building thermal capacity – measured in hours – and short-term dynamics associated with the HVAC system and the control system – measured in minutes. Thus we can then investigate all aspects of dynamic control strategies (Hartman 1988, Athienitis 1988) that seek to improve whole building performance by exploiting the interaction between the building thermal mass and the HVAC system.

Park *et al.* (1989) concluded that their choice of initial conditions for the simulations affected the results significantly, and that the most difficult task was the preparation of input data for a simulation.

An alternative approach to complete simulation of whole building systems is to focus on specific interactions between the building, its HVAC system and the control system through frequency response studies, and with transient analysis (Athienitis *et al.* 1990). This can be achieved by means of frequency domain techniques in conjunction with Laplace transforms. Building Laplace transfer functions, which typically represent the effect of heat sources or temperature changes on room temperatures, may be employed for three types of studies:

- **Frequency domain analysis:** the building significant thermal characteristics can be identified, and various design options may be compared on a relative basis though frequency domain studies of the magnitude and phase variation of the building transfer functions; this technique is particularly suitable for passive solar analysis (Mathews and Richard 1989).
- **Thermal control studies:** these studies are performed by first combining the building s-domain transfer functions with the HVAC system and control Laplace transfer functions. After determining the overall transfer functions through standard block diagram algebra techniques we may perform two types of analysis: (a) transient control studies such as the system response to setpoint and load changes (these are performed through numerical inversion of Laplace transforms); and (b) frequency response analysis of the open-loop and closed-loop transfer functions for stability studies (Athienitis *et al.* 1987).
- **Building heating/cooling load and temperature calculations:** these are performed by means of discrete Fourier series. The building transfer functions are calculated at discrete frequencies ($s = j\omega$). A discrete Fourier transform (DFT) of the weather data is performed and the time domain load or room temperature variation is obtained through complex algebra techniques and an inverse discrete Fourier transform (IDFT). Moreover, z-transfer functions, readily obtainable from the s-transfer functions, may be used for energy analysis.

Design of thermal control systems and operation strategies for buildings has often been based on frequency response techniques which employ simple building thermal models (Fisk 1981), and thus simplified transient analysis (Zhang and Warren 1988) As reported by Borresen (1981), the use of oversimplified models for rooms in selecting control parameters results in throttle temperature ranges and integral times in proportional integral control that are larger than necessary.

The method described uses detailed building thermal network models to determine the Laplace transfer functions that are employed for both energy calculations and thermal control analysis. A unified approach to building energy analysis and building thermal dynamics/control studies is proposed. By using such an approach, the thermal processes in a building and the dynamic interactions with an HVAC system can be modelled in detail, and with high accuracy. The building, its HVAC system and the control system are considered as one thermal dynamic system with all components represented by Laplace transfer functions.

8.2.2 Methodology

A building is represented by a thermal network, and transfer functions are obtained by performing an energy balance at all nodes in the Laplace domain. Both lumped and distributed elements can be considered with this approach. Simple models that do not represent in detail infrared radiation heat exchanges between room interior surfaces can usually be solved analytically (e.g. Section 5.1). For such cases the building transfer functions are determined in symbolic form as a continuous function of the Laplace domain variable s; s is equal to $j\omega$ in the frequency domain. For detailed thermal network models an analytical solution is usually not feasible. The admittance matrix that arises from the energy balances of the network is then inverted at discrete frequencies and the transfer functions are thus obtained only at these frequencies. The analytical function is obtained by fitting a polynomial $N(s)/D(s)$ through a modified least squares technique. Analytical transfer functions (Athienitis *et al.* 1987, Fisk 1981) may be used for parametric analysis in order to examine for effects of variables such as thermal mass on the thermal lag time (related to phase angle of the relevant transfer function) associated with radiant gains. Analytical s-transfer functions are also required for transient control studies in which a numerical inversion technique for Laplace transforms is employed.

BUILDING TRANSFER FUNCTIONS

Building transfer functions generally provide the response of interest – heat flow or temperature for unit heat input or unit temperature change at a node in the thermal network. The most important transfer function required in the present method is the impedance transfer function

$$Z(i,j) = T(i)/Q(j) \tag{8.12}$$

which represents the temperature change for node *i* due to unit heat input at node *j* for a given frequency. Thus, for heat input $Q(j)$ the room temperature change $T(1)$ (1 = room air node) is equal to

$$T(1) = Z(1,j) \times Q(j)$$

It is often useful to determine a transfer function not only for individual room temperatures, but also for an effective room temperature such as the operative temperature (ASHRAE 1997). The operative temperature is defined as the uniform temperature of an enclosure in which an occupant would exchange the same amount of heat by radiation plus convection as in the actual non-uniform environment; it is given by

$$T_e = (h_r T_{mr} + h_c T_{ai})/(h_r + h_c) \tag{8.13}$$

where T_{ai} is the air temperature, T_{mr} is the mean radiant temperature, and h_r and h_c are radiative and convective coefficients, respectively, for a person or object (sensor). The operative temperature transfer functions $X(i)$ are given by

$$X(i) = T_e/Q(i) \tag{8.14}$$

and represent the effect of a source $Q(i)$ acting at node i on the operative temperature.

The thermal network model for a typical room over a basement, with one window and convective auxiliary heating, is depicted in Figure 8.2. The thermally massive walls are modelled by a two-port distributed element, while the room air and lightweight room contents are modelled by a lumped thermal capacitance. Although this capacitance has no effect on load calculations because of the relatively low frequencies involved, it is important to include it for short-term (high-frequency) control studies. Each two-port element in Figure 8.2 represents the equivalent two-port for each wall, obtained after multiplying the cascade matrices for each massive and non-massive layer. The resistances connecting node 1 (room air) to the interior surfaces represent convective conductances given by

$$U_{1j} = \mathrm{A}(j)h_c(j) \tag{8.15a}$$

The radiation conductances interconnecting room interior surface nodes 2–8 are given by

$$U_{ij} = A(i)\sigma 4T_m^{\,3}F^*_{ij} \tag{8.15b}$$

where σ is the Stefan–Boltzman constant and $4T_m^{\,3}$ is a linearization factor which is based on an estimated mean temperature, T_m. The radiation exchange factors

(a)

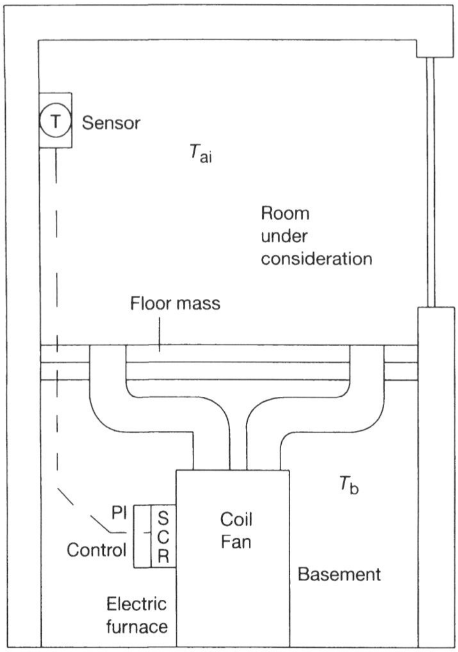

Figure 8.2 (a) Room with convective heating; (b) thermal network for top zone. Nodes: 1 = room air, 2–8 are room interior surfaces – 2 = window, 3 = front wall, 4 = ceiling, 5 = right wall, 6 = back wall, 7 = floor, 8 = left wall

(b)

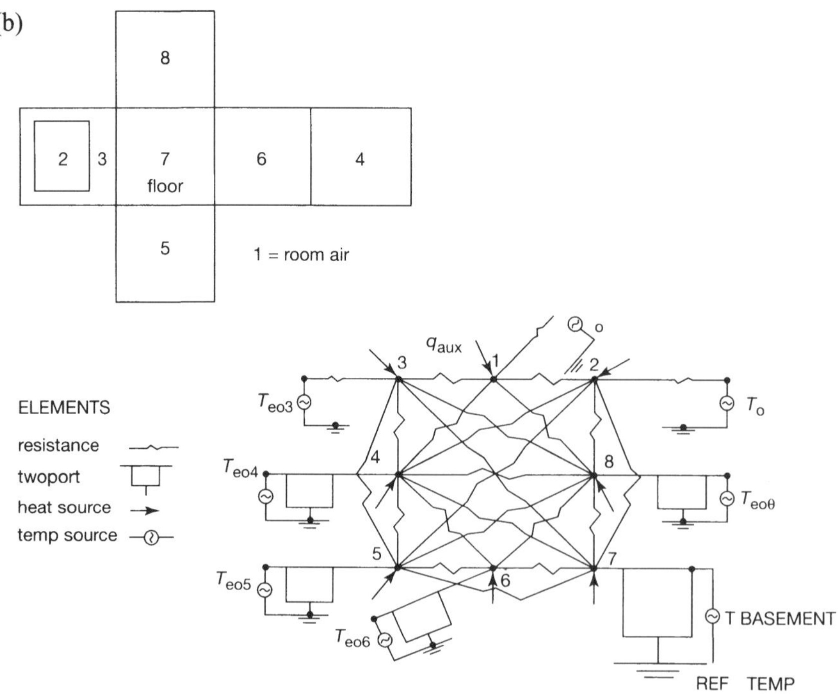

(F^*_{ij}) between the pair of surfaces under consideration (i and j) are determined from the radiation view factors, F_{ij} (also denoted by $F(i,j)$), and the radiative properties of the room surfaces as follows:

$$F^*(i,j) = m(i,j)\varepsilon_i\varepsilon_j/\rho_i \tag{8.15c}$$

where $[m] = [M]^{-1}$; the elements of matrix $[M]$ are: $M(i,j) = I(i,j) - \rho_i F(i,j)$, with $I(i,j) = 1$ if $i = j$; otherwise, $I(i,j) = 0$ (identity matrix).

The energy balances at the room interior nodes for both models are readily obtained after replacing each wall by its Norton equivalent subnetwork (see Chapter 5) which consists of an equivalent heat source Q_{sc} and a self-admittance Y_{eq}, thereby eliminating all exterior nodes without discretizing the massive elements. The equivalent source Q_{sc} is equal to the wall transfer admittance times at external specified temperature. For the floor with self admittance Y_{fs} and transfer admittance Y_{ft} we have $Q_{sc} = -T_b Y_{ft}$ (the negative sign is because of the sign convention used).

The energy balance for the model (with summations ΣU_{ij} over $j = 1$–8) is as follows:

$$\begin{bmatrix} sC_a+\Sigma U1J+U_{inf} & -U12 & -U13 & -U14 & -U15 & -U16 & -U17 & -U18 \\ -U12 & Y_{2s}+\Sigma U2J & -U23 & -U24 & -U25 & -U26 & -U27 & -U28 \\ -U13 & -U23 & Y_{3s}+\Sigma U3J & -U34 & -U35 & -U36 & -U37 & -U38 \\ -U14 & -U24 & -U34 & Y_{4s}+\Sigma U4J & -U45 & -U46 & -U47 & -U48 \\ -U15 & -U25 & -U35 & -U45 & Y_{5s}+\Sigma U5J & -U56 & -U57 & -U58 \\ -U16 & -U26 & -U36 & -U46 & -U56 & Y_{6s}+\Sigma U6J & -U67 & -U68 \\ -U17 & -U27 & -U37 & -U47 & -U57 & -U67 & Y_{7s}+\Sigma U7J & -U78 \\ -U18 & -U28 & -U38 & -U48 & -U58 & -U68 & -U78 & Y_{8s}+\Sigma U8J \end{bmatrix} \begin{bmatrix} T1 \\ T2 \\ T3 \\ T4 \\ T5 \\ T6 \\ T7 \\ T8 \end{bmatrix} = \begin{bmatrix} Q1 \\ Q2 \\ Q3 \\ Q4 \\ Q5 \\ Q6 \\ Q7 \\ Q8 \end{bmatrix}$$

$$[Y]_{N \times N}\{T\}_N = \{Q\}_N \tag{8.16}$$

The elements of the admittance matrix may be obtained by inspection from the topology of the network. The diagonal entry $Y(i,i)$ is equal to the sum of the component admittances connected to node i. Off-diagonal entry $Y(i,j)$ is the sum of component admittances/conductances connected between nodes i and j, multiplied by –1. The heat source vector element $Q(i)$ is the sum of the heat sources connected to node i (positive if directed to the node). As can be seen, for thermal networks the admittance matrix has certain important characteristics: (1) it is symmetric, with all off-diagonal elements $Y(i,j)$ being real, and equal to –(conductance U_{ij}); (2) all capacitances and all self-admittances appear in the diagonal entries which are consequently complex. Note that the diagonal element $Y(i,i)$ also contains the lumped capacitance for air C_a.The transfer functions of interest are the elements of the inverse of $[Y]$, that is the impedance transfer functions $Z(i,j)$. The temperature of node i for each frequency is given by:

$$T(I) = \sum_{j=1}^{8} Z(i,j)Q(j) \tag{8.17}$$

The functions $Z(i,j)$ are determined at specific frequencies ($s = j\omega_n$) by inverting the admittance matrix $[Y]$. The operative temperature, T_e, is a scalar function of nodal temperatures $\{T(1)...T(8)\}$.

Operative temperature transfer functions, $X(i)$: the key to accurately determining the weighting factors for the individual surface temperatures in T_e is to utilize a detailed model for radiant exchanges and for T_{mr}, which is obtained through the appropriate radiative energy balances. In the derivation below, the sensor is assumed to sense the same operative temperature as a person. The long-wave radiant heat gain, q_s, sensed at a location in the room is given by

$$q_s = A_s\varepsilon_s\,\{[q^+_i F(s,i) + ... + q^+_I F(s,I)] - \sigma T_s^4\} \tag{8.18}$$

where A_s is the effective sensing area, q^+ represents radiosity, $i = 1,...,I$ (surface number), and T_s is the temperature of the sensor (or object/person). The mean radiant temperature, T_{mr}, is defined by the equivalent energy balance:

$$q_s = A_s\varepsilon_s\sigma(T_{mr}^4 - T_s^4) \tag{8.19}$$

For simplicity the sensor or person is assumed to be located in the centre of the room, and the view factors $F(s,i)$ between the sensor and surface i are approximated as equal to A_i/A_t (A_t = room area). If the effect of location within the room is to be studied the factors $F(s,i)$ must be calculated accurately. The surface radiosities, q^+_i, are given by the matrix equation (Edwards 1981):

$$\{q^+\} = [M]^{-1}\{q_e\} \tag{8.20}$$

where $q_e(j) = \varepsilon_j T_j^4$.

Note that the inverse $[M]^{-1}$ is already obtained in evaluating the exchange factors F^*_{ij} (equation (8.15b)), so no new matrix inversion is required. The long-wave emissivities, ε_j, of the surfaces can usually be approximated as $\varepsilon = 0.9$. After substituting in equation (8.18) for q^+, then for $F(s,i)$ equal to A_i/A_t and linearizing the T^4 terms, q_s may be expressed with the following matrix equation:

$$q_s = A_s h_r[(\varepsilon/A_t)\{A\}'[M]^{-1}\{T\} - T_s] \tag{8.21}$$

where $h_r = \varepsilon_s 4\sigma T_m^3$, and $\{A\}' = [A_1,...,A_i,.A_I]$ (transpose of the area vector).

After linearizing equation (8.19) and comparing it with equation (8.21) we may deduce that T_{mr} is given by

$$T_{mr} = \{D_{mr}\}'\{T\}$$

with row vector $\{D_{mr}\}'$ given by

$$\{D_{mr}\}' = (\varepsilon/A_t)\{A\}'[M]^{-1} \tag{8.22}$$

($'$ = transpose).

The transfer functions $X(i)$, which denote the effect of source elements $Q(i)$ on the operative temperature, will now be derived. First, rewrite equation (8.13) as a function of room interior temperatures in matrix form,

$$T_e = \{D\}'\{T\} \tag{8.23}$$

where $\{D\}' = [D(1),D(2),...D(8)]$, while $\{T\}$ is the room interior temperature vector. The weighting factors, $D(i)$, are given as follows. The weight, $D(1)$, for the room air temperature is equal to $h_c/(h_r + h_c)$, as can be seen from equations (8.13) and (8.23). The entries $D(2)...D(8)$ are equal to the relative weighting factors of the surface temperatures in the mean radiant temperature times the ratio $h_r/(h_r + h_c)$; they are therefore determined by multiplying vector $\{D_{mr}\}$ of equation (8.22) by $h_r/(h_r + h_c)$. Substituting for $\{T\}$ from equation (8.23) we may deduce that

$$T_e = \{D\}'[Y]^{-1}\{Q\} \tag{8.24}$$

Now define an adjoint vector $\{X\}$ by

$$\{X\}' = \{D\}'[Y]^{-1} \tag{8.25a}$$

Multiplying both sides of equation (8.25a) by $[Y]$, we obtain

$$\{X\}'[Y] = \{D\}' \tag{8.25b}$$

Now, taking the transpose of both sides of equation (8.25a),

$$[Y]'\{X\} = \{D\}$$

but $[Y]' = [Y]$, therefore,

$$[Y]\{X\} = \{D\} \tag{8.25c}$$

and

$$\{X\} = [Y]^{-1}\{D\} \tag{8.25d}$$

Since $\{D\}$ is a function of room geometry and surface properties only, it must be evaluated only once for a particular room. Further, since $[Y]$ is a function of room construction and frequency only, it has to be determined only once for each frequency. Thus, equation (8.25d) needs to be solved only once for each frequency to obtain the operative temperature transfer functions $\{X\}$.

8.2.2 Analysis of building transfer functions

Substantial insight into building thermal behaviour may be obtained by studying the magnitude and phase angle of the important transfer functions. Consider for example the transfer functions Z11 and Z17 in the detailed model which represent the effects of heat sources at node 1 (room air) and 7 (floor) respectively on the temperature of node 1:

$$Z11(s) = \left.\frac{T1(s)}{Q1(s)}\right|_{\text{all other sources set to zero}} \tag{8.26a}$$

$$Z17(s) = \left.\frac{T1(s)}{Q7(s)}\right|_{\text{all other sources set to zero}} \tag{8.26b}$$

where s is the Laplace domain variable.

The magnitude of Z17(jω) may be used to determine the approximate room temperature swings due to solar radiation absorbed at the floor surface as follows.

If S_7 represents solar radiation absorbed at the floor interior surface and $|S_7(j\omega_1)|$ represents the magnitude of its fundamental harmonic, the approximate temperature swing amplitude is given by $|Z17(j\omega_1)| \times |S_7(j\omega_1)|$. Perhaps a more significant result is the time delay between the peak of $S(t)$ (noon for south-facing windows) and the resulting peak of the room temperature; this is approximately equal to φ_{17}/ω_1 (seconds) where φ_{17} is the phase angle of Z17 ($\varphi_{17} = \tan^{-1}$ Im(Z17)/Re(Z17)).

Results. The room considered in the example is 7.3 m wide by 2.4 m high and the north–south depth is 6.7 m. The south-facing double-glazed window area is 11.1 m^2. The thermal mass is 4 cm thick concrete on the floor with thermal conductivity equal to 1.8 W/m^2 K, density 2242 kg/m^3 and specific heat capacity 840 J/kg K. The interior lining on vertical walls and on the ceiling is a 13 mm thick gypsum board. The insulation is 3.5 RSI on vertical walls, 7.4 RSI on the ceiling and 1 RSI on the floor, which connects to a basement.

Figure 8.3 depicts the magnitude and phase variation (Bode plots) for Z11 with a carpeted floor versus a concrete floor (4 cm thick). The magnitudes are non-dimensionalized by dividing by the steady-state ($\omega = 0$) values of the transfer functions. Examination of Figure 8.3 indicates that the room response can be approximately separated into a short-term dynamics high-frequency range and a low-frequency range. The separation between short-term and long-term building thermal dynamics begins at frequencies of approximately 35 cycles per day (cpd) or periods of 40 minutes. In the high-frequency, short-term dynamics region, the room's air thermal capacitance is significant and the difference between Z11 for the concrete floor and the carpeted one is small in both phase and magnitude: that is, the effect of thermal mass is minimal in this region. More extensive studies (Stylianou 1989) for different constructions have produced similar results for Z11 which represent the effects of convective heat gains or losses. Short-term dynamics are particularly important for feedback control studies. For lower frequencies such as one cycle per day the magnitude and phase of Z11 is a strong function of room construction.

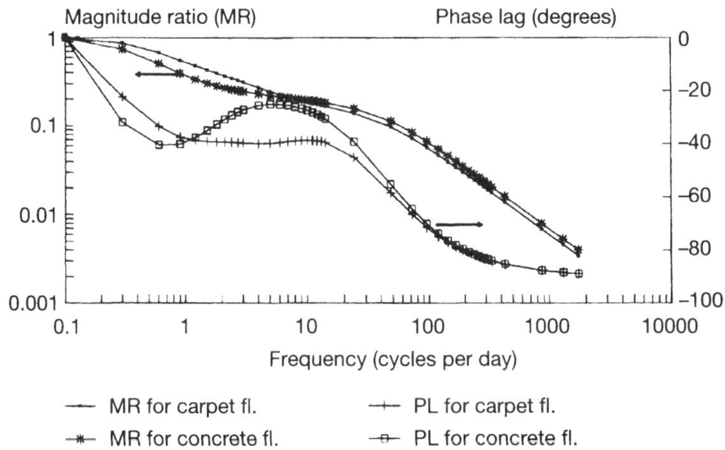

Figure 8.3 Magnitude and phase (Bode) plot of transfer function Z11 for a room with two different floor inner linings: 4-cm thick concrete; carpet. (Magnitudes are divided by Z11(*C*=0)

Examination of Figure 8.4 shows that for Z17, which represents the effect of radiant gains at the floor on room air temperature, there is no clear separation between short- and long-term dynamics. There is a significant difference between the response of the room with the carpeted floor and that with the concrete floor for all frequencies. The thermal lag time is significantly larger for the concrete floor. As expected, this indicates that feedback control of systems with high radiant loads or radiant heating systems is more complex due to the larger thermal lag times involved. Therefore, transfer functions such as Z17 need to be determined accurately for such applications.

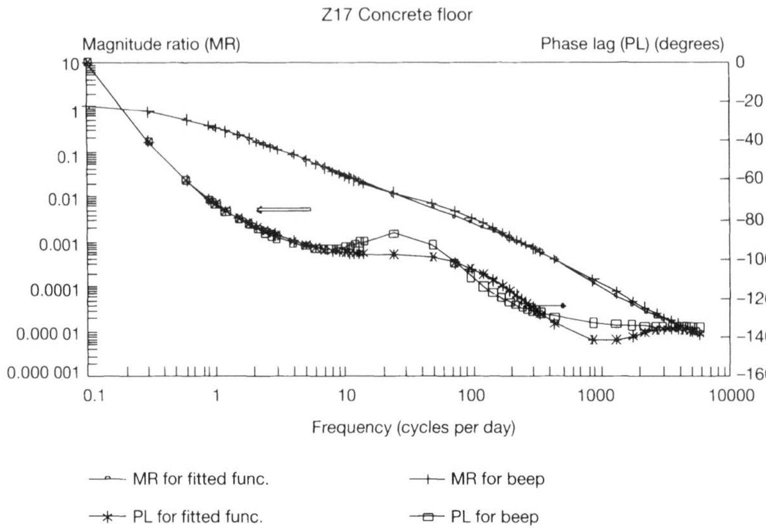

Figure 8.4 Exact discrete frequency responses and fitted third-order transfer function for Z17

8.2.3 Determination of s-transfer functions from discrete responses

The transfer functions in Figure 8.3 were obtained at the discrete frequencies shown through inversion of the admittance matrix [*Y*] defined by equation (8.16). The objective now is to synthesize transfer functions $Z_f(s)$ as ratios of complex polynomials $N(s)/D(s)$ from the discrete frequency response data. They are determined as a continuous function of s by performing a least-squares complex interpolation on the discrete values by means of a technique developed by Levy (1959), appropriately modified for the present application. The method is briefly described next. Let $Z(\mathrm{j}\omega)$ be the actual discrete frequency responses (generated by BEEP, Building Energy and Enviroment Program) which may be represented by the sum of its real and imaginary parts:

$$Z(\mathrm{j}\omega) = R(\omega) + \mathrm{j}I(\omega) \tag{8.27a}$$

The fitted polynomial function $Z_f(s) = N(s)/D(s)$ may be expressed at the given frequencies as:

$$Z_f(\mathrm{j}\omega) = \frac{A_0 + A_1(\mathrm{j}\omega) + A_2(\mathrm{j}\omega)^2 + \cdots + A_m(\mathrm{j}\omega)^m}{B_0 + B_1(\mathrm{j}\omega) + \cdots + B_n(\mathrm{j}\omega)^n} \tag{8.27b}$$

Examination of equation (8.27a) reveals that $A_0 = Z(\omega = 0)$ and $B_0 = 1$. Because the steady-state value of Z, that is $Z(\omega = 0)$, is known, A_0 does not need to be evaluated. Separating the numerator and denominator into real and imaginary parts and using a least-squares technique (Athienitis *et al.* 1990), we may obtain a satisfactory third-order transfer function for Z11 and Z17.

Typical results obtained with the technique are given in Figure 8.4 for Z17 which shows both magnitude and phase variation of the fitted and exact transfer functions over the frequencies of interest, and the magnitudes are non-dimensionalized by dividing by the steady-state ($\omega = 0$) values of the transfer functions. The resulting magnitude ratio was determined with an accuracy of 1%.

8.2.4 Thermal control studies

The building, HVAC system and control transfer functions are combined through block diagram algebra to obtain the overall system s-transfer functions. These transfer functions are then studied in the frequency domain ($s = j\omega$) for stability and other analyses or are used for transient analysis of overall system response to setpoint and load variations using numerical inversion of Laplace transforms. This technique is described and illustrated through examples in the next section.

Consider a room with convective heating, simple feedback control and N input-loads $Q(J)$ influencing room temperature $T(1)$ (also denoted T_{ai}) through transfer functions $Z(I,J)$ (see Figure 8.5) or $X(J)$ if we employ a sensor which detects operative temperature – the ideal variable to control. The resulting room air temperature variation as a function of setpoint and load changes is then obtained through block diagram algebra as

$$T_{ai}(s) = T_{sp}\frac{G_f G_c G_p Z11}{1 + G_f G_c G_p Z11\, G_s} + \sum_{J=1}^{N}\left\{Q(J)\frac{Z(I,J)}{1 + G_f G_c G_p Z11\, G_s}\right\} \tag{8.28}$$

where T_{sp} is the setpoint and G_f, G_c, G_p (equal to $G_{p1} \times G_{p2}$) and G_s are the transfer functions of the final control element, controller, heating system, and sensor respectively. All variables are a function of s. $G_f G_c G_p Z11 G_s$ is the loop transfer function that may be used for stability analysis using the Nyquist criterion or the Bode criterion (Stephanopoulos 1984). For example, the Bode criterion may be used to determine the period and gain at the stability limit, known as the ultimate period and ultimate gain K_u respectively, at which the phase angle of the loop transfer function (for proportional control) is equal to 180°. The selection of the actual gain is then based on various tuning techniques such as the Ziegler–Nichols method (1942). Normally tuning is performed on site for HVAC systems because the system parameters are usually not accurately known; self-tuning adaptive algorithms may also be used. However, the analysis is useful for comparison of

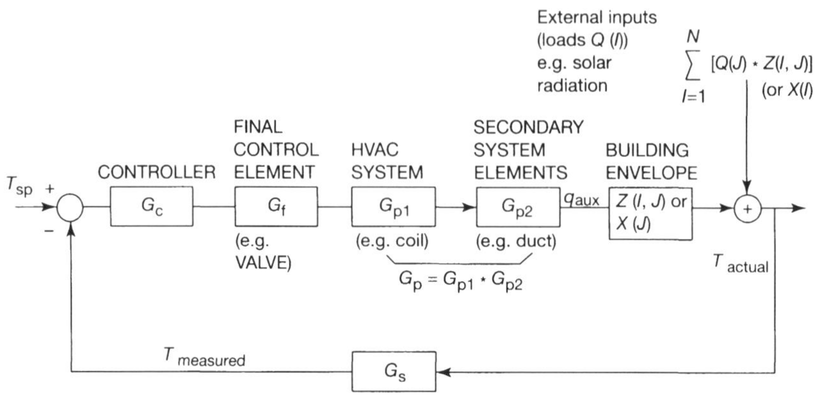

NOTE: 1. External load $Q(J)$ acts at node J and affects room air temperature through transfer function $Z(1,I)$, or operative temperature T_e through transfer function $X(J)$.

2. In some systems (e.g. electric home heating systems) the transfer functions G_{p1} and G_{p2} may be expressed as one combined heating system transfer function Gp for convenient analysis.

3. The auxiliary heating/cooling q_{aux} is equivalent to a source $Q(J)$ applied at node J (J=1 for convective systems:– $T(1)$ = room air temperature; J=4 for ceiling heating; – $T(4)$= ceiling temp.; J=7 for floor heating; – $T(7)$ = floor temp.)

Figure 8.5 Block diagram for thermal control analysis

alternative control algorithms, the effects of sensors with different time constants, and of system parameters such as coil time constant and amount of thermal mass. After the control parameters are determined for the cases of interest, the system transient response to setpoint and load changes can be investigated using numerical inversion of Laplace transforms. Such studies also greatly assist the on-site tuning process.

For radiant heating analysis the transfer function Z11 in the first term of equation (8.28) is replaced by $Z1j$, where j is equal to the node number corresponding to the heated surface (in the model shown in Figure 8.2(b), j = 4 for ceiling heating). Moreover, the controlled temperature may not be the air temperature, but another temperature such as the operative temperature. In such a case the operative temperature transfer functions $X(I)$ are used instead of $Z(I,J)$ in equation (8.28).

8.2.5 Transient simulation with numerical inverse Laplace transforms

Different numerical methods for Laplace transform inversion are available, but in general, they require the determination of poles and residues which might be computationally prohibitive if multiple poles occur. The method employed for numerical inversion of Laplace transforms such as $T_{ai}(s)$ given by equation (8.28) has been developed and applied to circuits by Vlach (1969) and Vlach and Singal (1983). This method provides an efficient way to calculate the time response of systems without determining poles or residues; that is, transfer functions that are transcendental and time delays may be included in the system. This permits the use of transfer functions for the building envelope which model the thermal mass as distributed elements. For numerical inversion, s is set equal to z_i/t, where t is the time and z_i are constants determined once and for all by Vlach and Singhal (1983) (Table 8.2). Thus, the overall closed-loop transfer functions, such as those defined by equation (8.24), do not need to be evaluated analytically but only at the selected time steps; for example, $s = z_i/t$ for one time step and $s = z_i/(t + dt)$ for the next. Thus the method can be easily incorporated in programs dealing with frequency response analysis of systems. Note that also for frequency domain studies loop transfer functions need only be evaluated at specific frequencies using complex algebra. Table 8.2 summarizes the numerical inversion of the Laplace domain response of T_{ai}.

Representative studies of room temperature response to step increases in heat gains will now be described (Figures 8.6 and 8.7) for the two different constructions in Figure 8.3, which represent different amounts of thermal mass. Figure 8.6 illustrates the effect of a radiative step heat input (1 kW) for the carpeted floor and for the concrete floor. It is determined by the inverse Laplace transform (ILT) of a step input times the relevant transfer function, i.e.

$$\text{Change of } T_{ai}(\mathrm{t}) = \mathrm{ILT}\{T_{ai}(\mathrm{s})\}$$

where

$$T_{ai}(\mathrm{s}) = 1000/(s\mathrm{Z}17_{f}(\mathrm{s}))$$

Figure 8.6 also shows the analytical result (denoted ANL), and excellent agreement is indicated. Figures 8.5–8.7 also show the equivalent first-order time constants obtained from the transient response as the time by which the temperature reaches 63% of its steady state value. As expected, because of its higher thermal mass, a concrete floor takes longer to reach steady state than a carpeted floor (10.75 versus 5.5 hours) and results in a lower temperature rise due to its larger thermal resistance. Thus, for radiative loads the effect of thermal mass on room temperature response is higher than for convective loads, and longer times are

Table 8.2 Numerical inversion of Laplace domain response (*z* are poles and *K′* residues)

$z_1 \equiv 11.830\,093\,739\,168\,19 + 1.593\,753\,005\,885\,813j$	$z_2 \equiv 11.220\,853\,779\,395\,19 + 4.792\,964\,167\,565\,670j$
$z_3 \equiv 9.933\,383\,722\,175\,002 + 8.033\,106\,334\,266\,296j$	$z_4 \equiv 7.781\,146\,264\,464\,616 + 11.368\,891\,649\,049\,93j$
$z_5 \equiv 4.234\,522\,494\,797\,000 + 14.957\,043\,781\,281\,56j$	$K'_1 \equiv 16\,286.623\,680\,504\,79 - 139\,074.711\,551\,605\,1j$
$K'_2 \equiv -28\,178.111\,713\,051\,63 + 74\,357.582\,372\,741\,76j$	$K'_3 \equiv 14\,629.740\,252\,331\,42 - 19\,181.808\,185\,018\,36j$
$K'_4 \equiv -2870.418\,161\,032\,078 + 1674.109\,484\,084\,304j$	$K'_5 \equiv 132.165\,941\,247\,487\,6 + 17.476\,747\,988\,771\,64j$

$p = 1, 2..80$ $\Delta t = 30$ s $t_p = p\Delta t$ time array $s_{i,p} = \dfrac{z_i}{t_p}$

The Laplace transfer function to be inverted is $V_{i,p} = T_{ai}\left(s_{i,p}\right)$

with initial condition: $T_{ai_0} = 0°C$

The time domain response is given by:

$$T_{ai_p} = \frac{1}{-t_p}\left[\sum_i \mathrm{Re}\left(K'_i V_{i,p}\right)\right]$$

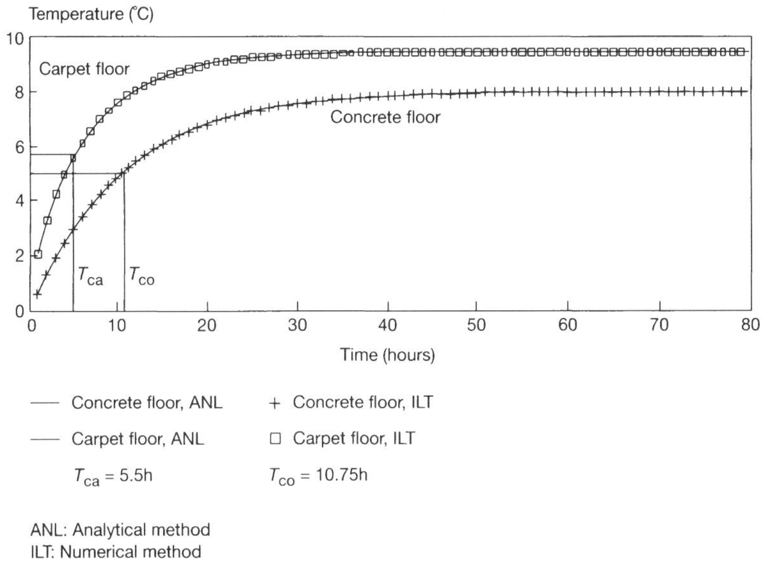

Figure 8.6 Room temperature response (change) to step input of convective heat gains (1 kW) for carpet floor and for concrete floor (4 cm) (ANL analytical, ILT numerical inverse Laplace transform)

required to achieve steady state. The effect of thermal mass thickness on room temperature response to radiative loads is shown in Figure 8.7; for 4 cm thick concrete floor lining the time constant is 10.8 hours, while for 6 cm thick concrete it is 14.5 hours.

The transfer functions describing the HVAC system components as well as the transfer functions describing the room model can easily be used to study their frequency responses by simply substituting s with jω and can be combined to produce the system's overall transfer function if needed, or can be studied individually or in subsystems in both the frequency and time domain.

This method does not require discretization of distributed elements such as a massive floor slab and also allows the use of dead times (time lags) without any artificial manipulation of the constants present, as is required if other numerical methods are used (Clark 1985). Further, it can be readily incorporated within a program for frequency response analysis because the transfer functions are already determined as functions of s (= jω).

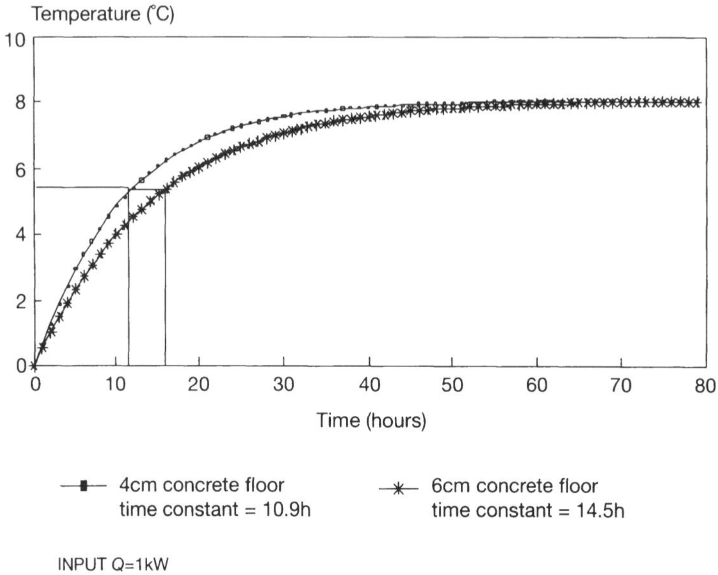

Figure 8.7 Effect of floor thermal mass (concrete) thickness on room temperature response to input of radiative gains (1 kW) at the floor surface (node 7)

8.2.6 Heating/cooling load and room temperature calculation

Heating/cooling load and associated room temperature calculations are performed with the same building transfer functions employed in frequency response and thermal control studies. These computations are performed by means of discrete Fourier series. The building transfer functions are calculated at discrete frequencies ($s = j\omega_n$) and a discrete Fourier transform (DFT) of the weather data is performed. The frequency domain response is then evaluated through complex algebra and the time domain load or room temperature variation is subsequently obtained through an inverse DFT as follows.

In frequency domain simulation periodic conditions are usually assumed; for example, if the simulation is to be performed for a week, it is assumed that all previous weeks have been the same. The steps needed for a periodic steady-state solution are as follows:

- Decide the number N of harmonics to be analysed for. If n represents a harmonic number and P is the time length of the simulation or analysis (e.g. a day or a week), then a harmonic's frequency ω_n is equal to $2\pi n/P$.
- Obtain the appropriate discrete Fourier series representations for the sources. An arbitrary source $M(t)$ is represented by a Fourier series (IDFT) of the form

$$M(t) = \sum_{n=-N}^{N} m_n(\mathrm{j}\omega)\exp(\mathrm{j}\omega_n t) \tag{8.29a}$$

the complex coefficients $m_n(\mathrm{j}\omega_n)$ being determined numerically by a DFT

$$m_n = \left[\sum_{k=1}^{K} M(t_k)\exp(-\mathrm{j}\omega_n t_k) \right] / P \tag{8.29b}$$

where $M(t_k)$ is the value of M at time t_k corresponding to point k (for a total of K values over the time length P).

- Determine the discrete frequency response $Z(\mathrm{j}\omega_n)$ of the output of interest to unit input at each node. The periodic response to each source is obtained by superposition of the output harmonics using complex (phasor) multiplication. For example, for an output temperature T(t), we have

$$T(t) = \sum_{n=-N}^{N} Z(\mathrm{j}\omega_n)m_n(\mathrm{j}\omega_n)\exp(\mathrm{j}\omega_n t) \tag{8.29c}$$

The total response to more than one input is determined by a double summation for all inputs $Q(i)$ and all frequencies of interest ω_n.
For example, for convective auxiliary heating we have

$$Q(1) = q_{\mathrm{aux}} + q_{\mathrm{int}} + q_{\mathrm{eq}} \tag{8.29d}$$

where q_{int} represents the convective portion of internal gains and q_{eq} is an equivalent source representing heat flow due to infiltration and is given by $U_{\mathrm{inf}}T_{\mathrm{o}}$. Therefore, by substituting equation (8.29d) in (8.16) and assuming that the room air temperature $T(1)$ is specified, we may determine the auxiliary heating/cooling power q_{aux} at each frequency of interest as:

$$q_{\mathrm{aux}}(\mathrm{j}\omega) = \left\{ T(1) - Z(1,1)[q_{\mathrm{int}} + U_{\mathrm{inf}}T_{\mathrm{o}}] - \sum_{\mathrm{j}=2}^{N} Z(I,J)Q(J) \right\} / Z(1,1) \tag{8.30}$$

where all quantities are evaluated as complex numbers for $s = \mathrm{j}\omega$ (N = number of nodes). Each source or specified temperature is represented by a discrete Fourier series (DFT) and the time domain variation of q_{aux} is obtained through an IDFT. For design day analysis 5–9 harmonics are usually adequate. These are the harmonics necessary for adequate representation of the inputs, that is, heat sources such as absorbed solar radiation, internal gains and ambient temperature T_{o}. One advantage of this approach over more commonly used methods is that the superposition principle is applied directly. Therefore, effects of various inputs may be studied separately, or a passive analysis ($q_{aux} = 0$) can be easily performed – in this case equation (8.29d) is directly applied. Note that the air temperature $T(1)$ in equation (8.30) can be a profile, that is, it may vary with time. Thus, optimum setpoint profile variations may be determined. The discrete Fourier series approach is described in more detail by Athienitis *et al.* (1987), including a model for a proportional control source in the thermal network and a technique for modelling time-varying parameters, such as a conductance representing infiltration based on the substitution network theorem.

8.3 Bibliography

ASHRAE (1989) *Ventilation for Acceptable Indoor Air Quality*. Standard 62-1989. ASHRAE, Atlanta, Georgia.

ASHRAE (1992) *Thermal Environmental Conditions for Human Occupancy*. Standard 55-1992, ASHRAE, Atlanta, Georgia.

ASHRAE (1995) *ASHRAE Handbook – Systems*, Atlanta, Georgia.

ASHRAE (1997) *ASHRAE Handbook – Fundamentals*, ASHRAE, Atlanta, Georgia.

Athienitis A.K. (1988) 'A predictive control algorithm for massive buildings', *ASHRAE Transactions*, **94**:2, 1050–1068.

Athienitis A.K. and Haghighat F. (1992) 'A study of the effects of solar radiation on the indoor environment', *ASHRAE Transactions*, **98**:1, 257–261.

Athienitis A.K., Sullivan H.F. and Hollands K.G.T. (1986) 'Analytical model, sensitivity analysis and algorithm for temperature swings in direct gain rooms', *Solar Energy*, **36**:4, 303–312.

Athienitis A.K., Sullivan H.F. and Hollands K.G.T. (1987) 'Discrete Fourier series models for building auxiliary energy loads based on network formulation techniques', *Solar Energy*, **39**:3, 203–210.

Athienitis A.K., Stylianou M. and Shou J. (1990) 'A methodology for building thermal dynamics studies and control applications', *ASHRAE Transactions*, **96**:2, 839–848.

Bhargava S.C., McQuiston F.C. and Zirkle L.D. (1975) 'Transfer functions for crossflow multirow heat exchangers', *ASHRAE Transactions*, **81**:2, 294–314.

Borresen B.A. (1981) 'Thermal room models for control analysis,' *ASHRAE Transactions*, **87**:2, 251–261.

Brandt S.G. (1986) 'Adaptive control implentation issues', *ASHRAE Transactions*, **92**:2, 211–219.

CIBSE (1986) *CIBSE Guide*, Volume A, CIBSE, London.

Clapp M.D. (1985) 'Cost savings using building management and control systems', *Proc. CLIMA 2000*, Vol. 3, Copenhagen, Denmark, pp. 33–36.

Clark D.R. (1985) *HVACSIM+ Building Systems and Equipment Simulation Program Reference Manual*, National Information Service, NBSIR 84-2996.

Edwards D.K. (1981) *Radiation Heat Transfer Notes*, Hemisphere Pub. Corp.

Fanger P.O. (1970) *Thermal Comfort*, McGraw-Hill, New York.

Fisk D.J. (1981) *Thermal Control of Buildings*, Applied Science Publishers, London.

Gagge A.P., Nishi Y. and Gonzalez R.R. (1973) 'Standard effective temperature – a single index of temperature sensation and thermal discomfort', Thermal comfort and moderate heat stress, Proceedings of the CIB Commission W45 (human requirements) Symposium held at the Building Research Station, London.

Gianccone A. and Gianfranco R. (1986) 'The influence of solar radiation distribution in a room on thermal comfort evaluation', *Proc. CIB*, 3185–3192.

Hartman T.B. (1988) 'Dynamic control: fundamentals and considerations', *ASHRAE Transactions*, **94**:1 599–609.

Hirsch J. (1985) 'Plan for the development of the next generation of building energy analysis computer software', *Proceedings of the Building Energy Simulation Conference*, Seattle, Washington, August, pp. 396–404.

ISO 7730 (1984) *Moderate Thermal Environments. Determination of the PMV and PPD Indices and Specifications of the Conditions for Thermal Comfort.* Geneva.

Kimura K. (1977) *Scientific Basis for Air-conditioning*, Applied Science Publishers, London.

Kelly G.E. (1988) 'Control system simulation in North America', *Energy and Buildings*, **10**, 193–202.

Levy E.C. (1959) 'Complex curve fitting', *IRE Transactions on Automatic Control*, May, 37–43.

Mathews E.H. and Richards P.G. (1989) 'A tool for predicting hourly air temperature and sensible energy loads in buildings at sketch design stage', *Energy and Buildings*, **14**, 61–80.

Olesen B.W., Rosendahl J., Kalisperis L.N., Summers L.H. and Steinman M. (1989) 'Methods for measuring and evaluating the thermal radiation in a room', *ASHRAE Transactions*, **95**:1, Paper no. CH-89-17-4.

Omega Engineering (1987) *Temperature Measurement Handbook*, Omega Engineering, Stanford, Connecticut.

Park C., Bushby S.T. and Kelly G.E. (1989) 'Simulation of a large office building system using the HVACSIM$^+$ program', *ASHRAE Transactions*, **95**:1, 642–651.

Park C., Clark D.R. and Kelly G.E. (1985) 'An overview of HVACSIM$^+$, a dynamic building/HVAC/control systems simulation program', *Proceedings of the Building Energy Simulation Conference*, pp. 175–185.

Shavit, G. and Brandt S.G. (1982) 'The dynamic performance of a discharge air-temperature system with a P-I controller', *ASHRAE Transactions*, **88**, 826–838.

Singhal K. and Vlach J. (1975) 'Computation of time domain response by numerical inversion of the Laplace transform', *Journal of the Franklin Institute*, **299**, 109–126.

Stephanopoulos G. (1984) *Chemical Process Control, An Introduction to Theory and Practice*, Prentice-Hall, Englewood Cliffs, New Jersey.

Stoecker W.F. *et al.* (1981) 'Reducing the peaks of internal air-conditioning loads by use of temperature swings', *ASHRAE Transactions*, **87**:2, 599–608.

Stylianou M. (1989) 'A computer method for building–HVAC system interaction and control applications', M.Eng. thesis, Centre for Building Studies, Concordia University, Montreal, Canada.

Thompson J.G. (1981) 'The effect of room and control systems dynamics on energy consumption', *ASHRAE Transactions*, **87**:2, 883–896.

Tobias J.R. (1973) 'Simplified transfer function for temperature response of fluids flowing through coils, pipes or ducts', *ASHRAE Transactions*, **79**:2, 19–22.

Vlach J. (1969) 'Numerical method for transient responses of linear networks with lumped, distributed or mixed parameters', *Journal of the Franklin Institute*, **288**, 99–113.

Vlach J. and Singhal K. (1983) *Computer Methods for Circuit Analysis and Design*, Van Nostrand Reinhold Co., New York.

Zhang, X. and Warren, M.L. (1988) 'Use of a general control simulation program' *ASHRAE Transactions*, **94**, Part 1, 1776–1791.

Ziegler J.G. and Nichols N.B. (1942) 'Optimal settings for automatic controllers', *ASME Transactions*, **64**, 759–765.

CHAPTER 9

Solar energy utilization techniques and systems

Several key modelling issues have been discussed in previous chapters concerning thermal analysis of solar buildings. This final chapter considers several integrative applications. Thermal analysis of a building with transparent insulation is presented in Section 9.1, followed by analysis of a solar building with phase-change thermal storage in Section 9.2. Section 9.3 describes a simple solarium model, Section 9.4 covers photovoltaic systems for buildings and Section 9.5 gives a description of recent advanced fenestration system applications.

9.1 Numerical model of a building with transparent insulation

9.1.1 Introduction

Transparent insulation (TI) used to replace conventional insulation in a solar collector wall has been proven to have a significant potential to reduce heating load in passive solar buildings (Twidell *et al.* 1994). Transparent insulation converts one or more walls of the building into solar collectors. Initially, a single glazing was used as the transparent cover of a wall. This was known as a Trombe wall system (Trombe *et al.* 1977). Cellular transparent materials were introduced to improve the solar transmittance and thermal resistance of the wall system (Olsen 1982, Goetzberger *et al.* 1983) after being applied to solar collectors (Francia 1961, Hollands 1965). These materials enhance thermal resistance, in addition to having high solar transmittance. Other types of TI are in use in the industry such as capillary glasses and aerogels. TI classes have been reviewed by Platzer (1994).

Use of TI with an appropriate amount of thermal storage achieves high solar energy utilization, especially for south-facing surfaces in winter. However, if too much solar radiation is absorbed, or if the thermal storage is inadequate, overheating problems will be experienced. There also several issues that need to be addressed during the design of a building with transparent insulation:

- The transparent insulation system should ideally be designed to satisfy the heating requirements on cold winter sunny days with solar energy, requiring practically no auxiliary heating.
- When more solar radiation is available than necessary it should be excluded through active means such as integrated motor-controlled blinds, or passive means such as thermotropic layers which change transmittance based on temperature (Georg *et al.* 1998).
- Appropriate design and control criteria need to be developed for the shading device to prevent overheating.
- While room air may not overheat, the room surface of the TI wall may exceed 30°C; this should be prevented by including an adequate amount of thermal storage mass and/or transporting some of the stored energy through forced or natural convection into the room.

The thermal lag effect in a building with a transparently insulated wall is more pronounced than in an ordinary building (Athienitis 1994, Ramadan and Athienitis 1998). The transfer admittance of a layer of concrete 20 cm thick, as considered in the TI wall of this section, introduces a time lag of about 5 hours in the room temperature rise; a further 3–5 hours is introduced by the rest of the room. Therefore, the maximum room temperature occurs between 8 p.m. and 11 p.m. Thus, it is necessary to determine the thermal response of a building with TI as a system. The fourth issue, while particularly important for practical implementation of TI into building systems, has not received much attention until now. It is considered in detail in this section. A thermal analysis simulation model of a building with one transparently insulated wall is described. It is employed to provide insight into issues of thermal comfort and energy conservation for a transparently insulated building during the design phase and to determine the effectiveness of various control strategies.

9.1.2 Transparent insulation system

A single zone building with a south-facing transparently insulated wall is modelled. The transparently insulated wall system, shown in Figure 9.1 (also considered in detail in Chapter 3), consists of an exterior single glazing protector, an air cavity, a transparent lexan honeycomb, a second air cavity, a concrete thermal storage layer, a third air cavity and an interior gypsum board layer.

The plastic transparent honeycomb material is assumed to be square cell. The cross-sectional area of the cell is determined to suppress convective heat transfer which may occur between the protector and the absorber prior to the introduction of the TI (Hollands *et al.* 1976). The aspect ratio of the cell, which is defined as the thickness of TI divided by the hydraulic diameter of the cell, must be high to reduce the conductive (air conduction) and radiative heat transfer of the TI layer (Tabor 1969). In addition, the thickness of the honeycomb wall material should be low to reduce the conductivity of the TI material (about 50 μm). However, with these thin materials, the aspect ratio is restricted to a value of around 5.

The *U*-value of the TI is assumed to be constant. For an inclined honeycomb layer, the dimensions of the cell may be calculated using the following equation (Hollands 1976):

$$A = f(\beta)(\mathrm{Ra}/2420)^{1/4} \quad (9.1)$$

with

$$f(\beta) = [4.45\cos(\beta - 60)]^{1/(11.52 - 6.56\sin\beta)}$$

for inclination angles 30–90° where A is the area of the honeycomb cell, β is the angle of inclination and *Ra* is the Rayleigh number (Holman 1986).

The thickness of the TI is chosen to preserve the high aspect ratio for the honeycomb cell to reduce radiative and conductive heat transfer. However, the material cost and transmission capability for thin honeycomb sheets restrict the value of the aspect ratio to about 5.

The air gap between the protector and the TI unit permits the use of a roller blind with low-emissivity reflective coatings; the blind enhances the thermal performance of the TI wall and excludes solar gains that would cause room overheating and discomfort. This is used to prevent overheating during days with excessive solar radiation. On winter nights the blind covers the facade to reduce

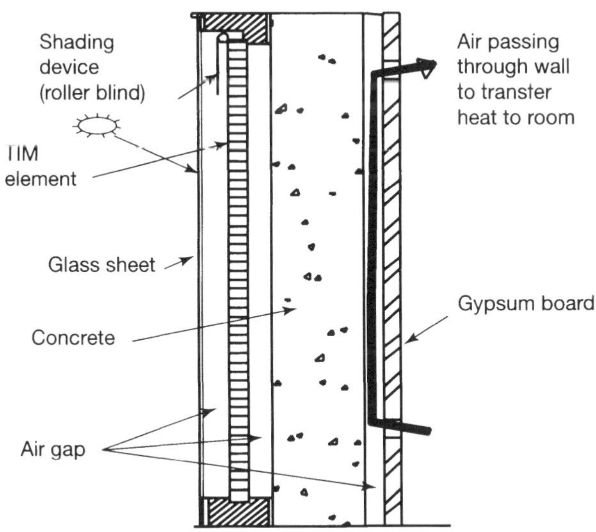

Figure 9.1 Cross-section of transparent insulation system and wall also showing how circulation can enhance heat transfer to room

long-wave radiation losses. However, the shading device is opened during summer nights to contribute to the reduction of the cooling load. The second air gap between the TI unit and the thermal mass decouples the radiative and conductive heat transfer of the transparent material. This air gap reduces the thermal conductance of the TI wall, especially when selective coatings cover the concrete layer (Hollands and Iyankaran 1985).

A third air gap is employed between the concrete thermal mass and the interior gypsum board layer. In addition to increasing the overall thermal resistance of the TI wall system, it prevents the inside surface temperature of the wall from rising too much and provides space for electrical wiring and other services. Thus, during clear winter days, the third air gap will aid in maintaining the inside wall temperature around comfort level. This space may also be coupled with room air and forced or natural ventilation may be employed to transfer some of the stored energy into the room air, thus preventing local overheating of the TI wall room surface, and improving utilization of solar gains. Forced convective heat transfer from the third air gap into the room is considered here.

The heat transfer through the air gaps is calculated as the combination of convective and radiative heat transfer based on relationships given by Elsherbiny *et al.* (1982).

The lexan transparent insulation used in the numerical simulations is square-celled honeycomb 48 mm wide with cell size 9.5 mm and wall thickness 0.076 mm. Various theoretical studies on the computation of solar transmittance of honeycomb transparent insulation used the Monte Carlo ray tracing technique. However, the complexity of this method has reduced its use. Hollands *et al.* (1978) has developed a simple approximate method to predict the incidence angle-dependent beam transmittance of honeycombs. This method considers refraction, scattering and absorption of radiation by the honeycomb walls. The sky and ground

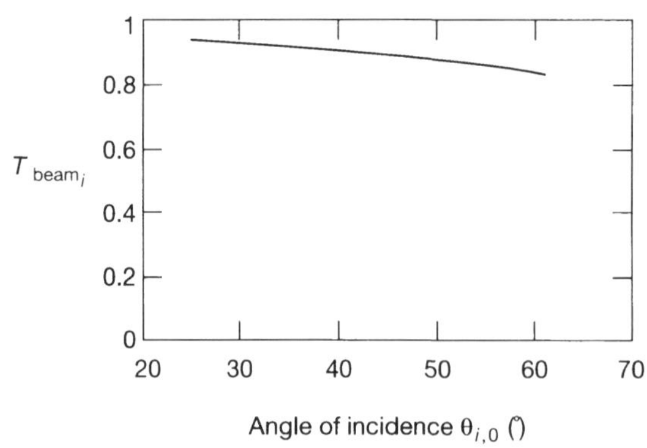

Figure 9.2 Beam transmittance of lexan honeycomb TI as a function of angle of incidence

diffuse solar transmittance of the TI is calculated as the transmittance for beam radiation at an equivalent incidence angle (Arulanantham and Kauskika 1994).

The solar diffuse radiation is assumed to be isotropic. Hence, the diffuse transmittance can be approximated as the transmittance of the same transparent material for beam radiation at an equivalent incidence angle. The diffuse transmittance is expressed in terms of sky and ground diffuse radiation (Brandemuehl and Beckman 1980). The incidence angle-dependent solar transmittance of the transparent honeycomb material is considered here based on these techniques. As can be seen from Figure 9.2, the beam transmittance is relatively high for most important incidence angles during the heating season. Figure 9.3 shows the total solar radiation transmitted through the TI system on a clear January day in Montreal. Hottel's clear sky model was employed (Hottel 1976).

The free convection heat transfer across a honeycomb can be characterized by a Nusselt number, *Nu*, which represents the ratio of the combined conductive and free convective heat transfer through the air to the corresponding heat transfer when the air is stagnant (Holman 1986). Hence, a Nusselt number of unity indicates conductive heat transfer and complete suppression of convective motion. On the other hand, as observed by Charters and Peterson (1972) through flow visualization tests, there is never a complete suppression of convection motion for an inclined cell. Thus, the Nusselt number is never exactly unity even at very small Raleigh number (*Ra*). However, the Nusselt number increases rather slowly with *Ra* until it reaches approximately 1.2, after which it rises rather steeply with *Ra* (Cane *et al.* 1977). Consequently, when $Nu = 1.2$ is reached, effective suppression is assumed to have taken place with purely conductive heat transfer of still air augmented by 20%.

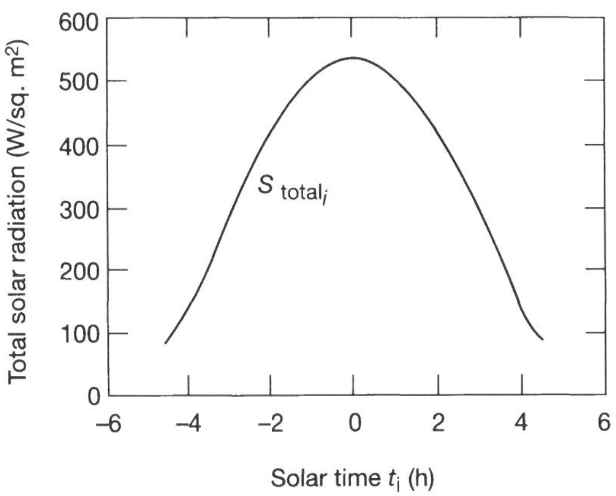

Figure 9.3 Total solar radiation transmitted through the transparent insulation system

Cane *et al.* (1977) reported an extensive experimental study on free convection across inclined square and hexagonal celled plastic honeycomb. They recommend the following relationship for Nusselt number:

$$Nu = 1 + 0.89\cos(\beta - 60)(Ra/2420A_{hc}^{4})^{2.88 - 1.64\sin\beta} \tag{9.2}$$

where β is the tilt angle (angle of inclination for honeycomb panel in degrees) and A_{hc} is the aspect ratio of the cell. This relation is valid when $30° \leq \beta \leq 90°$ and $Ra/A_{hc}^{4} \leq 6000$.

Although based on data for $A_{hc} \geq 4$, this relation is approximately valid for $A_{hc} = 3$ as well (Cane *et al.* 1977).

9.1.3 Simulation model

A numerical simulation model is presented in this section. The thermal performance of a test-room (dimensions $3 \times 3 \times 3$ m^3) with a south-facing transparently insulated wall is examined under different control strategies. The total thermal resistance of the walls is assumed to be 3.2 m^2/W K. The thermal mass in the TI wall is 20 cm of concrete. There is also 5 cm of concrete on the floor (top layer) and gypsum board on all vertical walls. The east wall has a double-glazed window with area equal to 2 m^2. The thermal behaviour of the room is simulated to evaluate whether any auxiliary heating is required for cold sunny days under different control strategies based on thermal comfort considerations. A cloudy day is also considered to evaluate the potential solar savings by comparison with a sunny day having the same mean temperature.

The explicit finite difference method is employed to solve the differential equation of heat transfer through the TI wall and between the other room components (walls, ceiling, floor and window). The thermal network of the TI wall and room is shown in Figure 9.4. The thermal mass (concrete layer) is discretized into five sublayers (control volumes); more layers did not significantly improve accuracy.

One thermal capacitance represents the thermal storage capacity of each wall interior layer. Using the explicit form of the finite difference approximation, an energy balance equation is applied for each node i as follows:

$$T(i,p+1) = \left(\frac{\Delta t}{C_i}\right)\left(q_i + \sum_j \frac{T(j,p) - T(i,p)}{R(i,j)}\right) + T(i,p) \tag{9.3}$$

where j represents all nodes connected to node i, q_i represents all heat sources at node i, p is the time interval and Δt is the time step, C_i is the thermal capacitance for node i and $R(i,j)$ is the conductive, radiative or convective thermal resistance between nodes i and j. An auxiliary heating system with proportional control is utilized and the room air temperature is permitted to fluctuate between 18°C and 24°C (up to 25°C is allowed).

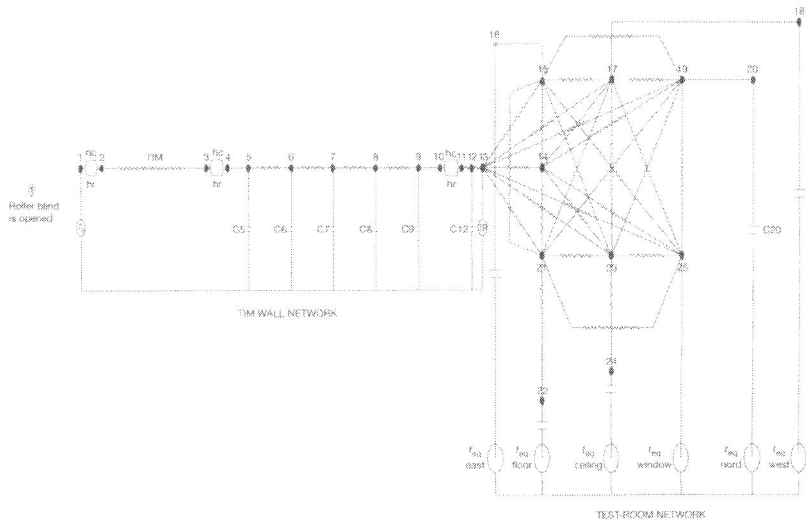

Figure 9.4 Thermal network of TI wall and room (node 14 is room air)

The TI wall is connected between nodes 1 (outside temperature) and 13 (inside surface of gypsum board). Node 14 represents room air, while nodes 15, 17, 19, 21, 23 and 25 represent the other room interior surfaces. Walls are connected to each other by surface nodes through a thermal radiative resistance and to the zone air node by a convective resistance. Both types of resistance are accurately calculated for each time step. Convective coefficients are calculated at each time step based on the following correlations recommended by ASHRAE (1989).

For horizontal air layer and upward flow (T_s = surface temperature):

$$h = 1.52(T_s - T_{air})^{1/3}$$

For horizontal air layer and upward flow (X = characteristic length):

$$h = 0.59[(T_s - T_{air})/X]^{1/4}$$

$$h = 1.31(T_s - T_{air})^{1/3} \tag{9.4}$$

The radiant exchange between the interior surfaces (assumed to be grey) may be calculated as follows (Holman 1986):

$$q_{i,j} = A_i F_{i,j} \sigma(T_i^4 - T_j^4) \tag{9.5}$$

where $F_{i,j}$ is the radiation exchange factor. (This factor includes the view factor and surface resistance of thermal radiation flow.)

The program developed accurately models the solar radiation on each face of the building. However, it does not model the actual distribution of the incoming solar radiation through the window to the interior surface of the zone. Instead, this distribution is approximated to be 70% for the floor and 30% for the other room surfaces.

Discretization or stability error results from the replacement of the differential equation by the explicit finite difference expression. This can be avoided by reducing the time step based on the following criterion (Holman 1986):

$$\Delta t \leq \min\left(\frac{C_i}{\sum_j \frac{1}{R(i,j)}}\right) \tag{9.6}$$

where the above summation is performed for any node with capacitance C_i connected to it; the summation considers all resistances connecting node i with neighbour nodes j. A time step of about 200 seconds was employed in most simulations.

To reduce the possibility of room overheating the blind is closed based on the following criteria:

- If $S < 100$ W/m^2 or $T_o > 14$°C where S is the total incident radiation and T_o is the ambient temperature (cloudy or warm weather).
- When $T_{13} > 29$°C or $T_{14} > 25$°C (where T_{13} is the the temperature of the surface of the TI wall facing the room and T_{14} is room air temperature (T_{ai})) to maintain good thermal comfort.

The last set of conditions is necessary and complementary as in certain cases the room temperature may be lower than 25°C but the TI wall surface temperature may exceed 29°C. Note that, owing to the thermal lag effect, T_{13} and T_{14} may continue to rise above 29°C and 25°C respectively even after the blind is shut. After several simulations, it was determined that if the second control measure above is applied, we would be excluding useful solar energy on several occasions; for example on several occasions $T_{14} < 20$°C (the setpoint) while $T_{13} > 29$°C and the blind shuts, excluding solar radiation while auxiliary heating is supplied. The best way to prevent this, and to reduce losses from the TI wall to the exterior environment while maintaining good thermal comfort, is to move air by natural or forced convection into the room. In this study, forced convection was assumed, with air flowing at 1 m/s through the cavity between the gypsum board and the concrete layer of the TI wall when $T_{13} > 24$°C if $T_{14} < 24$°C. Note that no air flow is necessary if the room temperature rises to 24°C. A heat transfer coefficient of about 10 W/m^2°C from the wall to the flowing air was estimated; this heat transfer is modelled with a negative heat source at node 12 and a positive source at node 14. The two heat sources are therefore given by

$$q_{12} = -A_{12}10(T_{12} - T_{14})$$

and

$$q_{14} = -q_{12} = A_{12}10(T_{12} - T_{14})$$

A useful temperature for evaluating thermal comfort is the operative temperature. This is defined as the uniform temperature of an enclosure in which an occupant would exchange the same amount of heat by radiation plus convection as in the actual non-uniform environment (see Section 8.1); it is given by:

$$T_{op} = (h_r T_{mr} + h_c T_{ai})/(h_r + h_c) \tag{9.7}$$

where T_{ai} is the air temperature, T_{mr} is the mean radiant temperature, and h_r and h_c are radiative and convective coefficients, respectively, for a person or object (sensor).

The model has been verified by comparison with experimental data (Ramadan 1998).

9.1.4 Simulations and results

An extreme cold day was initially considered for Montreal, Canada, latitude 45°N in January with the following sinusoidal outdoor temperature variation:

$$T_o = T_{om} + \Delta T_o \cos(wt - 5\pi/4) \tag{9.8}$$

where

T_{om} = mean (average) outside temperature = –15°C
ΔT_o = amplitude of T_o = 5°C
$w = 2\pi/(86\ 400\ \text{s})$

The auxiliary heating system is assumed to have a capacity of 2000 W. It is operated under proportional control; the room air is permitted to fluctuate between 18°C and 24°C ($q_{aux} = K_p(T_{14} - 20)$). Simulations were performed first to evaluate the potential energy savings due to the TI wall by comparing the energy consumption on a sunny cold January day with that on a cloudy cold day.

Figure 9.5 shows the room air temperature variation on two consecutive days after steady periodic conditions are achieved (5–8 days). For this sunny cold period (minimum outside temperature –20°C) the auxiliary heating required was 7.8 MJ per day without air flow through the TI wall and without blind control based on

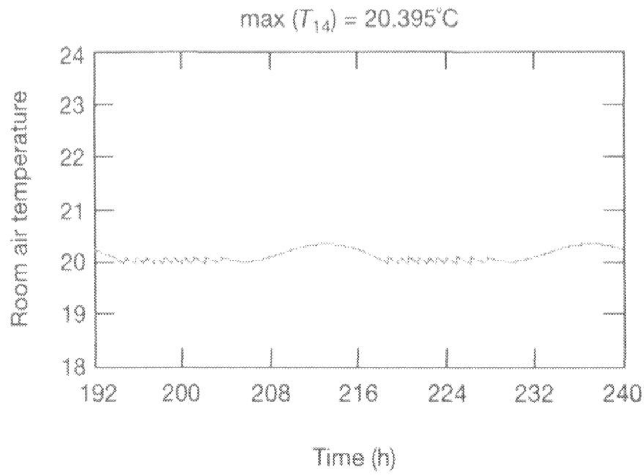

Figure 9.5 Variation of room air temperature on two consecutive sunny days (mean ambient temperature –15°C, setpoint 20°C). Q_h = 7.8 MJ (without control of T_{13} and no air flow through TI wall)

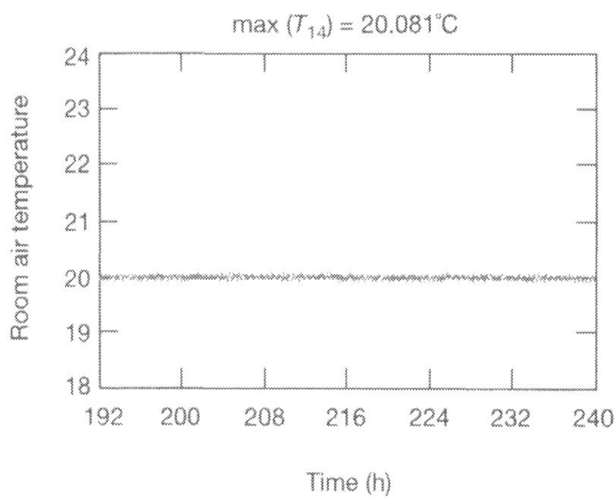

Figure 9.6 Variation of room air temperature on two consecutive cloudy cold days (mean ambient temperature –15°C, setpoint 20°C). Q_h = 72.5 MJ

T_{13} (blind control based on the other criteria), but T_{13} approached 32°C. Figure 9.6 shows the air temperature profile for the same outside temperature but cloudy conditions; in this case the energy consumption was 72.5 MJ, that is, the TI wall reduced energy consumption by almost 90% without forced airflow through the third cavity into the room. Further examination of Figure 9.5 reveals that the maximum air temperature occurs at approximately 9 p.m. due to the thermal lag effect of the TI wall (about 5 hours) and the rest of the room (between 3 and 4 hours) (Athienitis 1994).

The room air temperature fluctuates very little in the two cases because of the low outside temperature. The simulation shown in Fig. 9.6 was repeated with shut-off of the blinds when $T_{13} > 29°C$. The resulting energy consumption rose to 37 MJ. The resulting temperature profiles for the TI wall are given in Figure 9.7 for several days. The increase in energy consumption is due to the fact that the blind is shut when $T_{13} > 29°C$, excluding useful solar energy to maintain thermal comfort. The simulation was repeated with airflow through the third cavity of the TI wall when $T_{13} > 24°C$ (and $T_{14} < 24°C$). The resulting air temperature, operative temperature (ASHRAE 1989) and T_{13} are given in Figure 9.7 for several days. The resulting energy consumption was zero, i.e. no auxiliary heating was required on this extreme cold day when warm air is transferred through the TI wall to the room interior. Thus, one may conclude that the TI wall is adequate for this room. The resulting passive (no heating, since it is not required) response in Figure 9.8 shows that the maximum air temperature is 22.45°C, while the maximum T_{13} is about 27°C. The operative temperature is lower than the air temperature owing to the lower mean radiant temperature resulting from the low exterior temperature.

Simulations were also performed for mild cold days with mean temperature of 0°C and amplitude 5°C. In this case the risk of overheating is high. Figure 9.9 shows the resulting room air and operative temperature profiles for several days when all

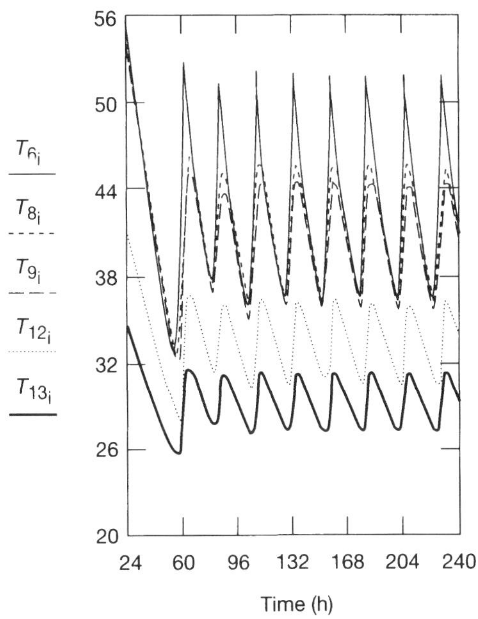

Figure 9.7 Temperatures in TI wall when outside temperature is –15°C (mean) for sunny conditions with blinds shut when $T_{13} > 29$ °C. $Q_h = 37$ MJ

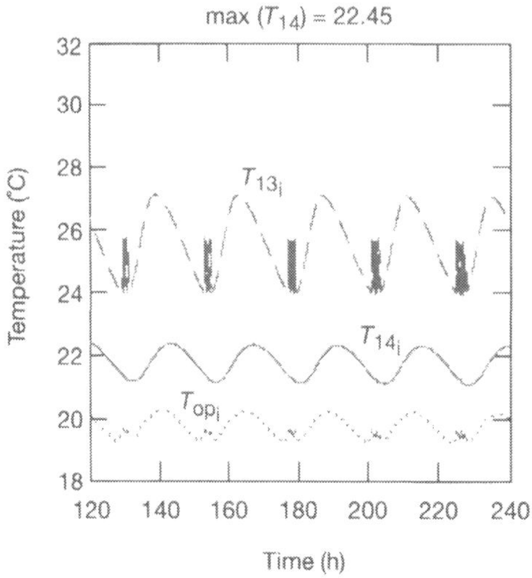

Figure 9.8 Room air temperature (T_{14}), operative temperature (T_{op}) and T_{13} (room facing temp. of TI wall) variation with blind shut when T_{13} > 29 °C or T_{14} > 25°C and air flow from TI wall to room when T_{13} > 24°C (and T_{14} < 24°C). Mean outdoor temp. –15°C, clear days (no heating required)

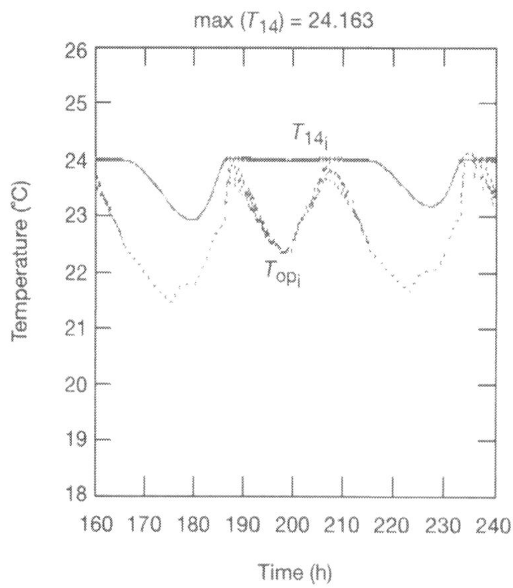

Figure 9.9 Room air temperature (T_{14}) and operative temperature (T_{op}) variation with blind shut when T_{13} > 29°C or T_{14} > 25°C and air flow from TI wall to room when T_{13} > 24°C. Mean outdoor temperature 0°C, clear days (no heating required)

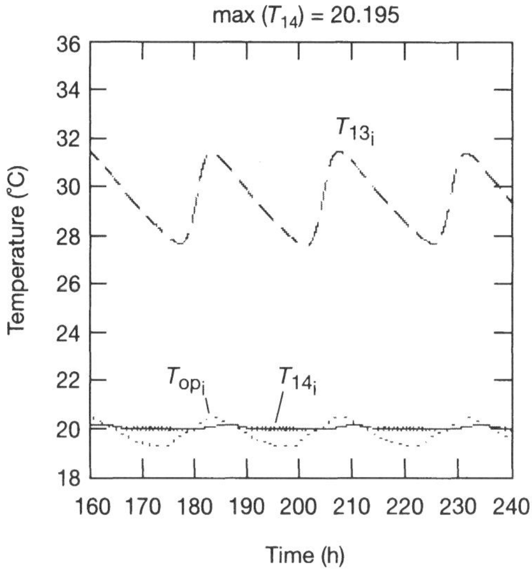

Figure 9.10 Temperature variations for same conditions as in Figure 9.9 but no air flow through TI wall into room (Q_h = 10 MJ)

control options are applied. No heating is required on a clear day and the maximum air temperature is about 24°C; however, the operative temperature, which is a more important comfort indicator, is for 90% of the time below 23°C. T_{13} reaches briefly a maximum of 31°C, i.e. higher than 29°C owing to the thermal lag effect. The simulation was repeated but with no air flow and Figure 9.10 shows the resulting temperature profiles. In this case, the room temperature remains around 20°C, but the daily heating energy consumption is 10 MJ and T_{13} reaches 31.5°C.

After comparing all the results shown above one may conclude that a south-facing transparently insulated wall may meet the heating energy requirements of an adjacent room on very cold sunny days. Appropriate control strategies for thermal comfort while maintaining the potential energy savings include shading (blind) control based on room temperature and TI wall temperature, as well as airflow through the TI wall into the room. Simultaneous implementation of these measures achieves the best energy performance and thermal comfort.

9.2 Thermal energy storage in solar buildings with phase-change materials

9.2.1 Introduction

Effective use of thermal storage in buildings will result in significant energy savings, but more importantly it will lower the peak demand for electrical power, thus reducing the need for construction of more power plants and contributing to a cleaner, less polluted, outdoor environment.

Energy demand from a system may be too high at times, and it may not coincide with availability. Thus, storage of energy is an effective way of reducing peak heating or cooling demand. Examples of possible use of thermal storage include the following:

- Storage of solar energy in the building envelope, or in an insulated enclosure such as a packed rockbed. Solar energy stored during the daytime is then used at night to reduce peak heating loads.
- Storage of heat or coolness generated electrically during off-peak hours for use during peak hours. Note that use of thermal storage, such as chilled water tanks, or ice storage, permits the installation of smaller chillers in large air-conditioning systems. The building envelope such as hollow-core floors may be utilized to store coolness (Zmeureanu and Fazio 1988).

HEAT STORAGE

Heat storage can be divided into two major areas, which reflect the time during which the heat is stored: short and long term (seasonal). These two categories reflect several periodicity patterns: diurnal (ambient air temperature, solar radiation, consumer habits), several days (typical weather period) and weekly (commercial electricity demand during weekends) for short-term storage. Long-term storage is mainly seasonal, such that heat produced during the summer can provide for heating demand during winter months. These categories can be subdivided according to the type of storage used. Although there are numerous ways to store heat, those that apply to space heating, that is, sensible, latent or chemical, are of the greatest interest.

BUILDING ENVELOPE MASS STORAGE

Thermal mass is used in the interior of buildings in order to reduce the room temperature swings caused by fast changes in the ambient temperature, by high solar gains, and by on/off cycling of a heater. Buildings with large glazed areas, such as office buildings or passive solar residential buildings, may present the most frequent discomfort problems during the heating season, when the solar radiation is utilized both for reducing the heating load and for natural daylighting. Maximum room temperatures as high as 32°C when the outside temperature is –15°C have been reported (Athienitis 1988). Of course, the problem can be eliminated, as in the summer, by preventing the solar radiation from entering the building through the use of shading devices or by opening the windows; this approach, however, defeats the use of windows as a means of reducing energy consumption. Another approach (Athienitis 1988) is to anticipate the room temperature rise and devise a control strategy for its prevention that makes more effective use of the storage capability of the thermal mass. A simple example of such a strategy is the use of a night setback in the thermostat setpoint. In a building with high interior solar gains, this leads to a faster cooldown of its thermal mass; thus, in the early morning, the mass, by being a couple of degrees lower in temperature than usual, may store more solar gains during the following daytime and reduce the room temperature swing.

Building envelope thermal mass reduces the room temperature swings caused by fast changes in the ambient temperature, by high solar gains, and by on/off cycling of heating/cooling systems. Buildings with large glazed areas, such as office buildings or passive solar residential buildings, may present the most frequent discomfort problems during the heating season, when the solar radiation is utilized both for reducing the heating load and for natural daylighting. Overheating is often observed in passive solar buildings during spring and autumn when some heating is required – usually in houses with large window areas with near-south (SE to SW) orientation and ineffective or inadequate thermal mass. During these shoulder periods, the solar gains are significant and usually exceed the instantaneous heating loads. Thus, significant amounts of thermal mass are required to store the excess solar gains without causing large room temperature swings. The use of phase-change materials (PCMs) in building products has rendered it feasible to store significant amounts of thermal energy in the building envelope without the uncomfortable temperature swings and large structural mass associated with sensible heat storage.

Storage as latent heat is caused by a phase change in the PCM. Energy storage may be achieved by melting the PCM and energy recovery by freezing it. This process involves the absorption or release of heat, even though little or no change in temperature occurs. PCMs may be incorporated in the building envelope to achieve latent heat storage. Feldman *et al.* (1989a, 1991) carried out extensive research on the use and stability of organic compounds for latent heat storage, including fatty acids (capric, lauric, palmitic and stearic), butyl stearate, dodecanol and polyethylene glycol 600. In addition to the studies of their properties, research was also carried out on materials that act as PCM absorbers. Various materials were considered, including different types of concrete and gypsum. The utilization of latent heat storage over a comfortable indoor temperature range in buildings can result in an increase in the thermal storage capacity in the range of 100–130% (Feldman *et al.* 1989b, Hawes 1991].

In the application described, PCM gypsum board was made by soaking conventional gypsum board in liquid butyl stearate (BS) (Feldman *et al.* 1989b), a PCM with phase change range of 16.0–20.3°C. The PCM gypsum board used contained about 25% by weight proportion of BS. Its thermal properties were measured with a differential scanning calorimeter (DSC).The objectives of this study were the following:

- Investigation of the thermal performance of PCM gypsum board used in a passive solar building.

- Estimation of the benefits from the application of PCM gypsum board in passive solar buildings in terms of the reduction of room overheating and energy savings.

Experiments in an outdoor test-room were performed and a mathematical model was developed to simulate the transient heat transfer process in the walls of the test-room which contain PCM gypsum board as inside lining.

9.2.2 Application

An outdoor test-room was used for this study (Athienitis and Shou 1991). The test-room, shown in Figure 9.11, is located in Montreal (45°N latitude). It has a double-glazed window facing 10° east of south. The room has three heating systems – baseboard, ceiling panels and floor heating – all electric and controlled by a computer that also controls the data acquisition system. The floor heating system was used. PCM gypsum board was attached over the existing drywall on the vertical walls of the test-room. Proportional control with a programmable setpoint was employed. The auxiliary heating power supplied is given by:

$$q_{\mathrm{aux}} = K_{\mathrm{p}}[T_{\mathrm{ai}}(t) - T_{\mathrm{sp}}(t)] \tag{9.9}$$

where $T_{\mathrm{ai}}(t)$ = measured room air temperature, °C
$T_{\mathrm{sp}}(t)$ = setpoint temperature at time t.

The experiments were conducted under winter weather conditions with outside temperature as low as –25°C. During the test, the room air temperature (or the globe temperature in some cases) was set to a high setpoint of 23°C during

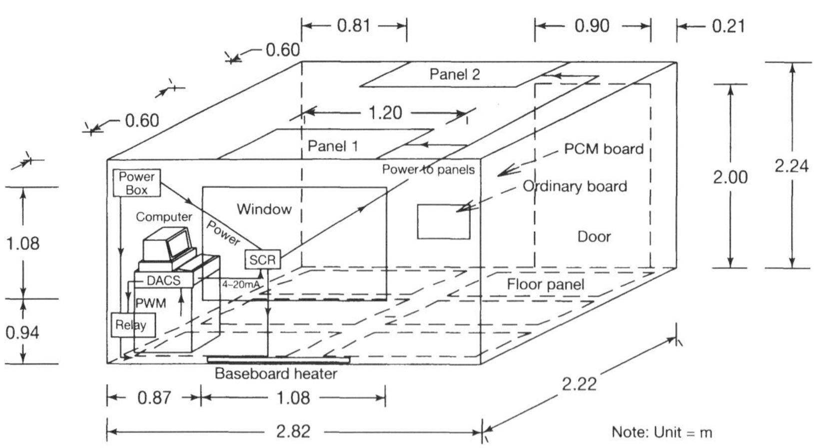

Figure 9.11 Schematic of outdoor test-room

the daytime (6 a.m. to 5 p.m) and 16°C during the rest of the time. When the room air temperature changes, the PCM in the gypsum board undergoes solid–liquid transition in the phase change range of BS (16–20.8°C). If the temperature rises through this range, the PCM wallboard acts as a thermal storage device. Conversely, if the temperature drops through this range, energy will be released.

Temperatures at approximately 30 different locations in the test-room were measured and recorded against real time. These included the front and back surface temperatures of the PCM gypsum board on each wall, the air temperature, the temperature of heating panels, etc. A 30 cm × 30 cm piece was cut out from the PCM gypsum board and replaced with ordinary gypsum board in order to compare the thermal behaviour of the two drywall specimens (with and without PCM) under similar thermal conditions. The front and back surface temperatures of the small piece of ordinary board were measured and recorded. The same measurements were taken on the PCM gypsum board immediately adjacent to this piece of ordinary gypsum board.

9.2.3 Mathematical model

The heat transfer mechanism in the PCM gypsum board is complex, especially when the butyl stearate is in the phase transition stage. During the freezing process, the PCM in the gypsum board could exist in three states (solid, liquid and two-phase). In addition, the physical and thermal properties of the gypsum matrix and the PCM in the gypsum pores are also different. To simplify the mathematical model, the following assumptions are made:

- The PCM and the gypsum matrix are considered as a body of uniform equivalent physical and thermal properties, such as specific heat, density, thermal conductivity and latent heat.
- The heat transfer process across the PCM gypsum board and the original wall is treated as a one-dimensional problem.

Based on the above assumptions, the wall heat transfer is treated as a one-dimensional, transient heat diffusion problem with uniform physical and thermal properties. The model takes into account the transient boundary conditions, absorbed solar radiation, the melting/freezing of the PCM and employs a non-linear film coefficient. The simulation of rate of latent heat release or absorbtion by the PCM is based on the test results from a DSC. The equivalent measured uniform properties of the PCM gypsum board were: a density of 900 kg/m^3, a specific heat capacity of 1260 J/kg°C, a thermal conductivity of 0.21 W/m°C and a latent heat equal to 30 700 J/kg.

The heat diffusion through the wall (Figure 9.12) is described by the following one-dimensional heat diffusion equation:

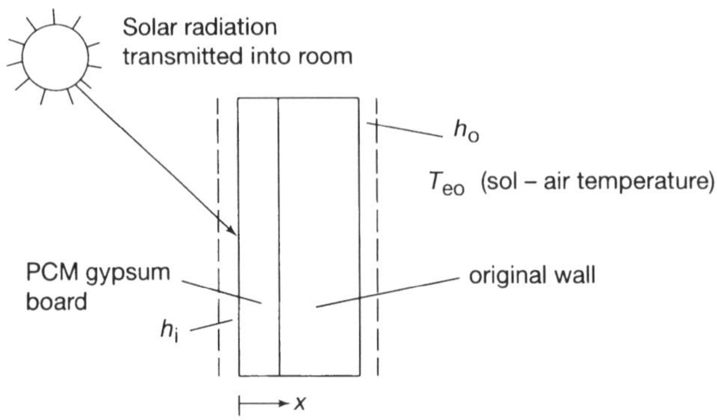

Figure 9.12 Wall section schematic for heat transfer model

$$\rho c \frac{\partial T(x,t)}{\partial t} = k \frac{\partial^2 T(x,t)}{\partial x^2} + q_l \tag{9.10}$$

with boundary conditions

$$-k \frac{\partial T(x,t)}{\partial x} = h_i(t)\left[T_{ai} - T(0,t)\right] + q_s(t) \qquad \text{at x = 0} \tag{9.10a}$$

$$-k \frac{\partial T(x,t)}{\partial x} = h_o\left[T(x,t) - T_{eo}\right] \qquad \text{at outside surface} \tag{9.10b}$$

where
ρ = density, kg/m^3
c = specific heat, J/(kg°C)
k = thermal conductivity, W/m°C
$h_i(t)$ = inside film coefficient, W/m^2°C
h_o = outside film coefficient, W/m^2°C
q_l = latent heat flux when phase change occurs, Wq_s
(t) = absorbed solar radiation, W/m^2°C
t = time, s
$T(x,t)$ = temperature, °C
T_{eo} = temperature, °C
x = distance, m

The latent heat flux, which represents equivalent internal heat generation due to freezing of PCM at a specific temperature, was modelled by the enthalpy method. The enthalpy method (Özisik and Uzzell 1979) simulates the heat released from solidification as a volumetric heat-generation term:

$$q_l = \rho L(\mathrm{d}f_s/\mathrm{d}t) \tag{9.11}$$

where f_s is the solid fraction in the two-phase region at the solidus front and L is the latent heat for complete phase transition.

The value of latent heat L and the rate of solidification df_s/dt in equation (9.11) were obtained from a DSC test. The DSC curve was modelled in the temperature range of interest as shown in Figure 9.13. In the figure, T_1 and T_2 indicate the temperatures at which freezing process starts and ends respectively; q_{max} indicates the maximum latent heat flow. The area of triangle $T_1T_2q_{max}$ is the value of latent heat for the phase transition range T_1 to T_2. With known heating or cooling rate, R, we have:

$$L_p = \tfrac{1}{2}q_{max}(T_2 - T_1)(1/R) \tag{9.12}$$

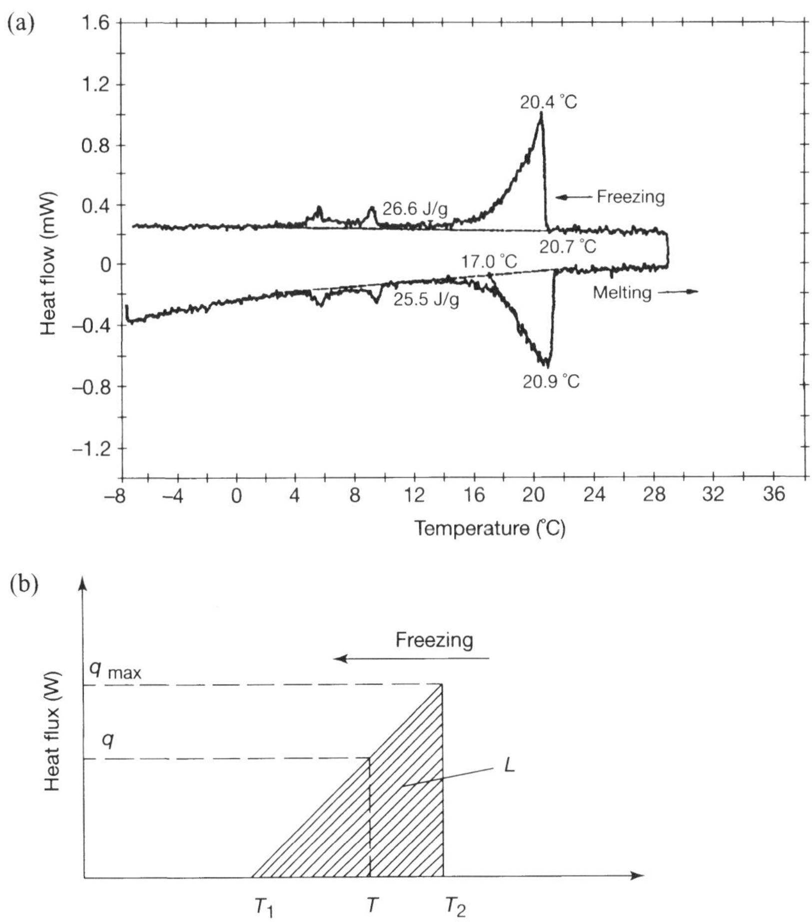

Figure 9.13 (a) Actual phase transition curve for PCM gypsum board from DSC; (b) approximation of DSC curve

where L_p is the latent heat for the temperature range T_1 to T_2, J/kg, R is the heating or cooling rate, °C/s, and T_1 and T_2 are the temperatures at which the freezing process starts and ends, respectively.

From trigonometry and interpolation, the latent heat flow, q_l, is thus determined at a temperature T as:

$$q_l = 2L_p R \frac{T - T_1}{(T_2 - T_1)^2} \tag{9.13}$$

Comparing the actual DSC curves with the approximation, it was found that the relationship between L_p and L is:

$$L_p = 0.62L \tag{9.14}$$

Numerical solutions have been successfully applied to modelling of building components with PCMs (Kedl 1991). The explicit finite difference method is particularly suitable for modelling non-linear heat diffusion problems such as the present case of heat transfer through the PCM gypsum board. It can easily accommodate non-linear heat transfer coefficients and temperature-dependent heat sources, such as those due to release of latent heat.

The finite difference thermal network model for the wall consisted of 14 nodes for the PCM board and 3 more nodes for the original wall behind it, as shown in Figure 9.14. The energy balance equation for the thermal network is:

$$T_{i,p+1} = \frac{\Delta t}{C_i}\left(q_i + \sum \frac{T_{j,p} - T_{i,p}}{R_{i,j}}\right) + T_{i,p} \tag{9.15}$$

The subscript i indicates the node for which the energy balance is written and j all nodes connected to node i, while p is the time interval; $R_{i,j}$ is the thermal resistance between nodes i and j, C_i is the thermal capacitance associated with node i, and q_i represents a heat source at node i such as absorbed solar radiation or internal heat generation (latent heat flow). The time interval Δt was selected based on stability considerations (see Chapter 4).

9.2.4 Results

Figure 9.15 shows representative experimental results for a sunny day. Temperatures of the front and back surfaces of the PCM drywall and the ordinary drywall cut-out are shown. As can be seen, the maximum front surface temperature of the PCM board is significantly lower (by 6°C) than the maximum temperature of the piece of ordinary drywall adjacent to it. This is due to the storage of solar gains as latent heat in the PCM board, while in the ordinary board the solar radiation is

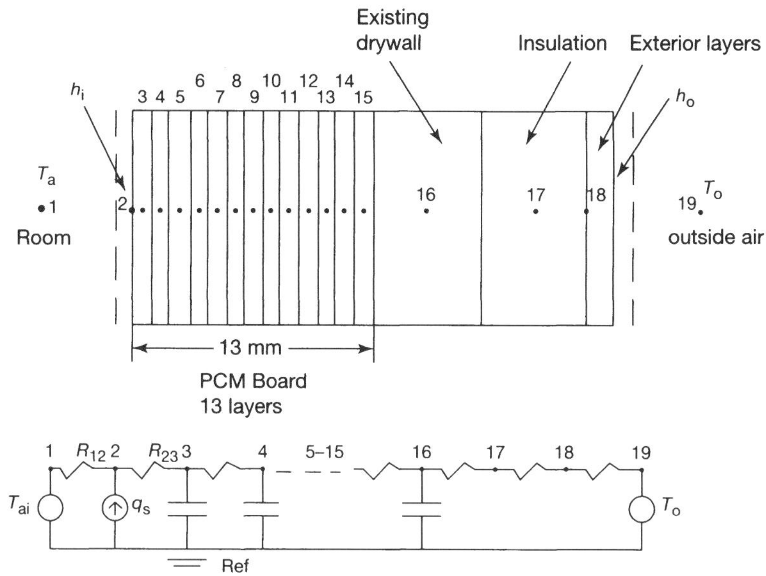

Figure 9.14 Wall section, nodal discretization and finite difference thermal network model

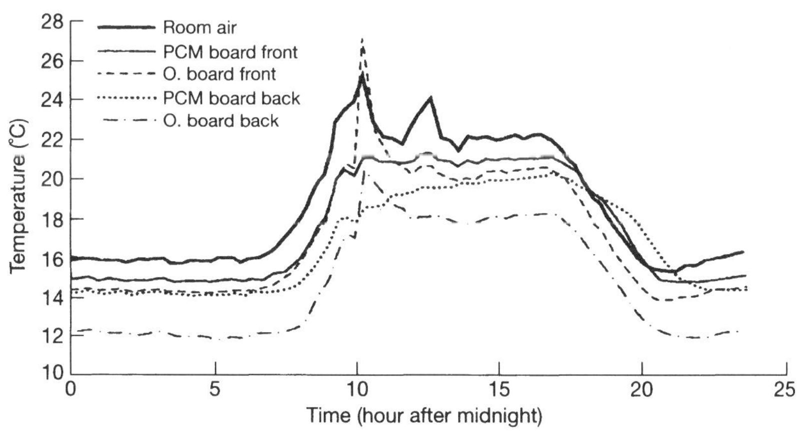

Figure 9.15 Experimental results from a typical winter sunny day

stored as sensible heat, resulting in an increase of surface temperature. The reduction of mean radiant temperature resulting from PCM use improves thermal comfort significantly, while permitting effective storage of solar gains.

In a previous study with the same test-room (Athienitis and Chen 1993) before PCM wallboard was employed, room air temperature was often as high as 30°C at noon on sunny days. Under similar circumstances such as Figure 9.15, the room air temperature was only 25–26°C when PCM wallboard was used. This decrease in maximum room temperature of about 4°C represents a significant improvement in thermal comfort. Therefore, a major conclusion is that the application of PCM gypsum board can significantly reduce overheating in a passive solar building.

Figure 9.16 compares experimental and simulation results for freezing of the PCM board in terms of front surface temperatures. The corresponding results for ordinary board (o. board) are also given. Generally, there is good agreement, with maximum error during the freezing process. This difference is primarily due to non-uniform phase change of the PCM in the drywall.

The above results show that during the freezing process which could last 7–11 hours, the surface temperature of PCM gypsum board is approximately 1–1.5°C higher than the surface temperature of ordinary board. With a total of about 20 m^2 PCM gypsum board used in the test-room, the increase in heat transferred from the wall to the room as compared to ordinary board (if it were used over the same area) is approximately 10 MJ. This is approximately equal to 15% of the total heating load, which indicates a significant potential for the reduction of energy consumption and peak loads.

The simulation results are generally in reasonably close agreement with the experimental results, indicating that the explicit one-dimensional non-linear finite difference model is satisfactory for simulating PCM gypsum board. The mathematical model may be used in conjunction with other building thermal analysis software to evaluate the design parameters and operational characteristics of

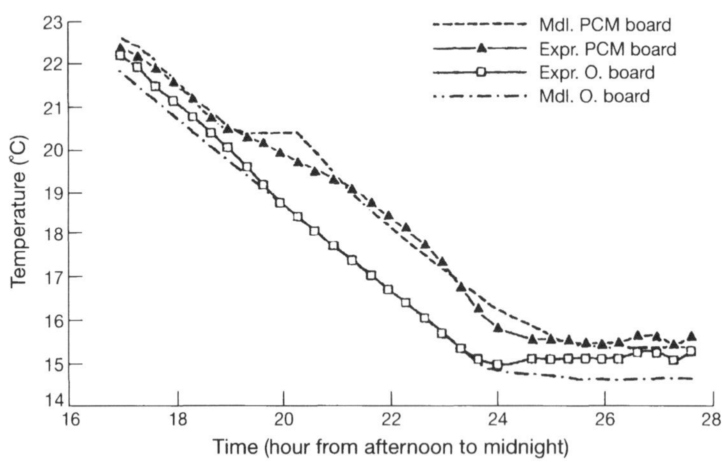

Figure 9.16 Comparison of experimental and theoretical results for freezing of the PCM (o. board = ordinary board)

buildings with PCM gypsum board or other phase change materials as inside wall lining.

9.3 Solarium design

This section discusses the thermal behaviour of a solarium and its optimum design and control. Recent developments in energy-efficient windows permit construction of solariums that can capture high amounts of solar gain while reducing transmission losses. Design guidelines for optimizing the design of a solarium in terms of shape, glazing type, orientation, thermal mass and integration of photovoltaic systems are discussed.

A solarium may be considered as a particular case of the direct gain passive solar system, but with most surfaces transparent, that is, made up of fenestration. Traditionally, a solarium, also often called a sunspace, is a space that the user will usually occupy during the daytime, enjoying the sunshine and doing mainly leisure activities. Thus, a small rise or decrease in temperature outside the comfort zone is acceptable because of the psychological benefits of the pleasant environment, which may also include plants; in some cases part of a solarium may act as a small greenhouse.

Solariums are becoming increasingly attractive both as a retrofit option for existing houses and as an integral part of new buildings. The major driving force for this growth is the development of new advanced energy-efficient glazing systems coupled with their drop in price. Low-emissivity double glazing with a thermal resistance close to 0.5 RSI is not significantly more expensive (10–15%) than ordinary double glazing (0.3 RSI) when the potential energy savings are considered. Reliable framing systems with thermal breaks that eliminate the thermal bridge effect are also readily available at relatively affordable prices. It is thus increasingly cost-effective to add a solarium to an existing house as a major renovation option to enhance the quality of the built environment and the value of the house. For example, consider a typical two-storey cottage built in Montreal with the rear of the house facing south or south-east. An appropriate location for a solarium would be the corner shown in Figure 9.17 because the exterior brick layer of the two walls may also function as thermal storage. Additional thermal storage will be added on the floor to reduce temperature swings, and provide better thermal comfort.

The design of a solarium requires several important decisions:

- The shape (dimensions L, W); it is best to make L/W as large as possible if L is the length of the surface facing south or near-south.
- The tilt of surface At is an important variable; in order to maximize net energy gains (solar gains minus transmission heat losses) a simulation for average days in each month must be performed determining this quantity.
- One design option is to integrate a photovoltaic system into one surface to

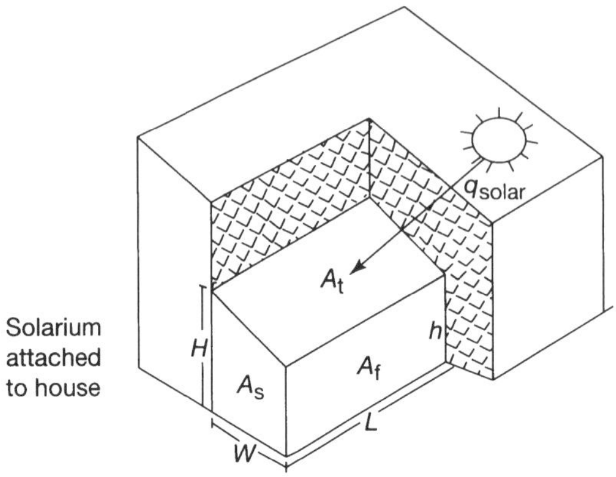

Figure 9.17 Schematic of solarium built as retrofit to existing house

generate also some electric energy while providing shading (Eicker and Fux 1999).

- Determination of the quantity of thermal storage required to prevent overheating of the solarium and to prolong the possible period of stay during the daytime.

Shading control measures are necessary to reduce the risk of overheating when only some or little solar radiation is needed, such as in September or in May. Retractable, motor-controlled reflective shades may be utilized between the glazings to allow the desired quantity of solar radiation based on the signal from a photosensor and a programmable controller. Because a solarium will most likely not be used on cold nights, even if its temperature is controlled to a low value such as 10°C, it can still act as a buffer zone for the adjacent rooms of the house, reducing their nocturnal heat losses.

9.3.1 Simple thermal network model of solarium

Consider the solarium shown in Figure 9.17. A simplified thermal network that models thermal storage in the walls and floor of the solarium with one lumped capacitance is shown in Figure 9.18. The main house temperature T_r will be assumed to be constant. The outdoor temperature T_o will be assumed to vary in a sinusoidal manner, as is very common in building thermal analysis.

*U*r3 represents the conductance between the mid-point of the thermal mass and the interior of the house; *U*31 is the conductance of half of the thermal mass; *U*12 is the film conductance (convective – radiative) between the mass surface in the solarium and the solarium air (or solarium 'effective temperature node'); *U*2o is the sum of the conductance of the exterior (glazed) surfaces of the solarium U_g and U_{inf} (the infiltration conductance).

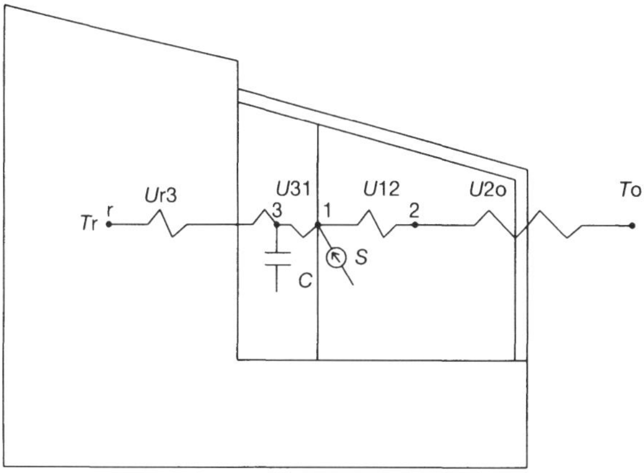

Figure 9.18 Simple thermal network model of solarium

We perform discretization in space and time; we represent the thermal mass with one capacitance and two resistances (T section); for more accuracy we may subdivide the mass into more layers (about one layer per inch) and an equal number of T sections. The thermal storage equation for a capacitance is

$$Q_{\text{stored}} = C\mathrm{d}T/\mathrm{d}t \tag{9.16}$$

The temporal derivative $\mathrm{d}T/\mathrm{d}t$ is approximated using explicit (forward) differencing as

$$\mathrm{d}T/\mathrm{d}t = (T_{p+1} - \mathrm{T}_p)/\mathrm{D}t$$

where Dt = time step, selected to prevent numerical errors.

9.3.2 Finite difference equations

Energy balance at node 3:

$$U3\mathrm{r}(T\mathrm{r}_p - T3_p) + U31(T1_p - T3_p) = C(T3_{p+1} - T3_p)/\mathrm{D}t$$

Energy balance at node 2:

$$U12(T1_p - T2_p) + U2\mathrm{o}(T\mathrm{o}_p - \mathrm{T}2_p) = 0$$

Energy balance at node 1:

$$U31(T3_p - T1_p) + U12(T2_p - T1_p) + S_p = 0$$

Rearranging:

$$T3_{p+1} = T3_p + [U3r(Tr_p - T3_p) + U31(T1_p - T3_p)]Dt/C \quad (9.17)$$

$$T2_p = (U12\,T1_p + U2o\,To_p)/(U12 + U2o)$$

$$T1_p = (U31\,T3_p + U12\,T2_p + S_p)/(U31 + U12)$$

After selecting initial conditions, the above equations may be solved for the temperatures at the next time step $p + 1$. Note that only the equation for node 3 involves a capacitance. The last two equations may be solved simultaneously, or a small time step Dt may be chosen and the left-hand side approximated by the values at $p + 1$. Therefore we can then solve all three equations marching forward in time from p to $p + 1$. The approximate solutions are given therefore by

$$T3_{p+1} = T3_p + [U3r(Tr_p - T3_p) + U31(T1_p - T3_p)]Dt/C$$

$$T2_{p+1} = (U12\,T1_p + U2o\,To_p)/(U12 + U2o)$$

$$T1_{p+1} = (U31\,T3_p + U12\,T2_p + S_p)/(U31 + U12)$$

Example

Assuming a clear day (1 February, latitude 45°N) and T_o

$$T_o = T_{om} + \Delta T_o \cos(\omega t - 5\pi/4)$$

where
T_{om} = mean (average) outside temperature = –10°C
ΔT_o = amplitude of T_o = 10°C,
$\omega = 2\pi/(86\ 400\ \text{s})$
Mass: 10 cm thick bricks on walls (k = 1.0 W/m K, p = 1500 kg/m^3, c = 800 J/kg°C). Determine the daily temperature variation with the finite difference method.

The result is shown in Figure 9.19.

For this case, no heating would be required during the daytime. Use of night insulation will also reduce nocturnal heat losses. Optimization of solarium dimensions may be achieved through performance analysis over typical days of each month in the heating season for several design options and selection of the optimum.

9.4 Photovoltaic systems for buildings

Photovoltaic (PV) systems found their first cost-effective applications to power remote buildings and devices located far from the existing electrical grid. They are

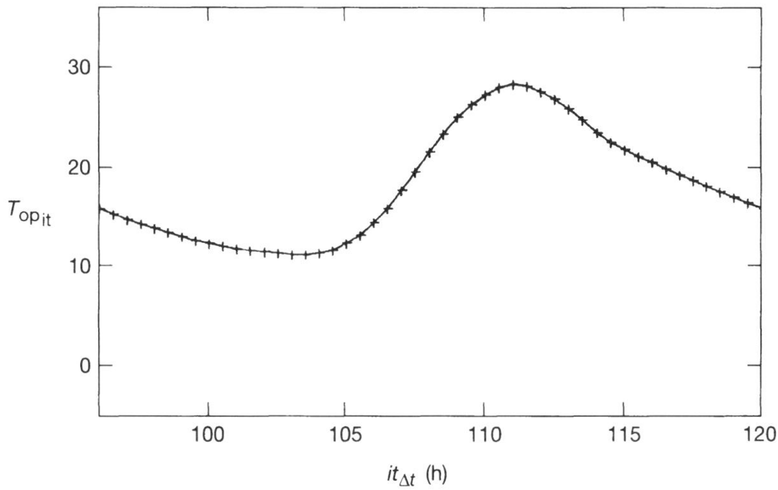

Figure 9.19 Operative temperature (T_{op}) on a clear winter (February) day

still a more expensive means of producing electricity than other energy sources. However, this is changing with their integration into the building envelope. When integrated into the building shell, such as the roof to replace ordinary shingles with photovoltaic shingles, or between two glazings in a semitransparent window, they perform a thermal function as well. Thus, their cost-effectiveness increases. Here we will consider first principles of PV systems design and typical building applications.

The central component of a photovoltaic system are the **photovoltaic (PV) modules** which generate electricity (direct current, DC) from sunlight. Other components are:

- **batteries (e.g. lead–acid)** which store the electric power generated and not immediately utilized
- **power conditioning circuitry such as battery charge regulators, DC–DC converters (to change DC output level), or DC–AC inverters which convert DC power to AC power**; they are essential in order to modify the power from the PV array and make it compatible with the current and voltage requirements of the load.

9.4.1 PV modules and solar cells

A PV module consists of solar (PV) cells permanently wired in series and/or parallel to form a single power unit, which is packaged for protection from the environment while permitting sunlight to reach the active surfaces of the cells. When solar radiation is incident on a semiconductor, the photons may transmit their energy to the valence electrons. Each time a photon breaks a bond an electron becomes free to move through the lattice. The absent electron leaves behind a vacancy (hole) that can also 'move' through the lattice as the electrons move around it. These holes can be considered like particles (positive). The movement of electrons and holes, if passed through an external circuit, can be just like any other current source.

A solar cell is made from a semiconductor material such as silicon or gallium arsenide which sometimes acts as conductor, other times as insulator. When solar radiation is absorbed it causes electrons to break loose from atoms, leaving atoms with a positive charge (hole). Under certain conditions the electrons and holes move in opposite directions, creating a current. In a solar cell the free electrons and hole move randomly and they eventually may recombine, or reach regions of the semiconductor containing different types of impurity, called **p-n junctions** (p indicates conduction by holes and n by electrons). When they reach the junction, they will move through an external circuit connected at the p-n junction, creating a current (see Figure 9.20a).

When a load is connected to an illuminated solar cell the current that flows is equal to the photogenerated current I_L due to generation of carriers by the sunlight minus the diode or dark current I_D due to recombination of carriers driven by the external voltage:

$$I = I_L - I_D$$

Thus, a PV cell is a **current source.** The main types of PV cells are **crystalline silicon, polycrystalline silicon cells** and **amorphous silicon** as well as **gallium arsenide, cadmium telluride**, etc. Crystalline cells are made from a single crystal (most expensive), polycrystalline cells from many small crystals (less expensive), while amorphous cells have no orderly atomic structure.

Crystalline silicon solar cells may be manufactured in single crystal wafers sown from boron-doped ingots grown from a melt, or in crystalline wafers cut from cast ingots. Wafers are usually sown to 10 cm $\times$ 10 cm size and 250–500 μm thick. Cells are produced by diffusing phosphorus into the p-type silicon wafer to make an n-type surface region. An electrical contact grid is deposited on the top face of the wafer exposed to sunlight and a solid contact on the other side (see Fig. 9.20a).

Amorphous silicon cells absorb incident solar radiation in a thin layer and they can thus be made thin; cells are made from layers of silicon about 1 μm thick deposited automatically on a substrate. The efficiency of conversion of sunlight to electricity is lower for amorphous cells (typically about 5–8%) than for crystalline cells (12–16%). (Note: laboratory efficiencies much higher have been obtained in 2001.) Amorphous silicon cells are typically used in applications such as calculators, but also increasingly in buildings such as for PV-shingles (cells integrated into roof shingles). For all cells, efficiency typically drops with rising operating temperature.

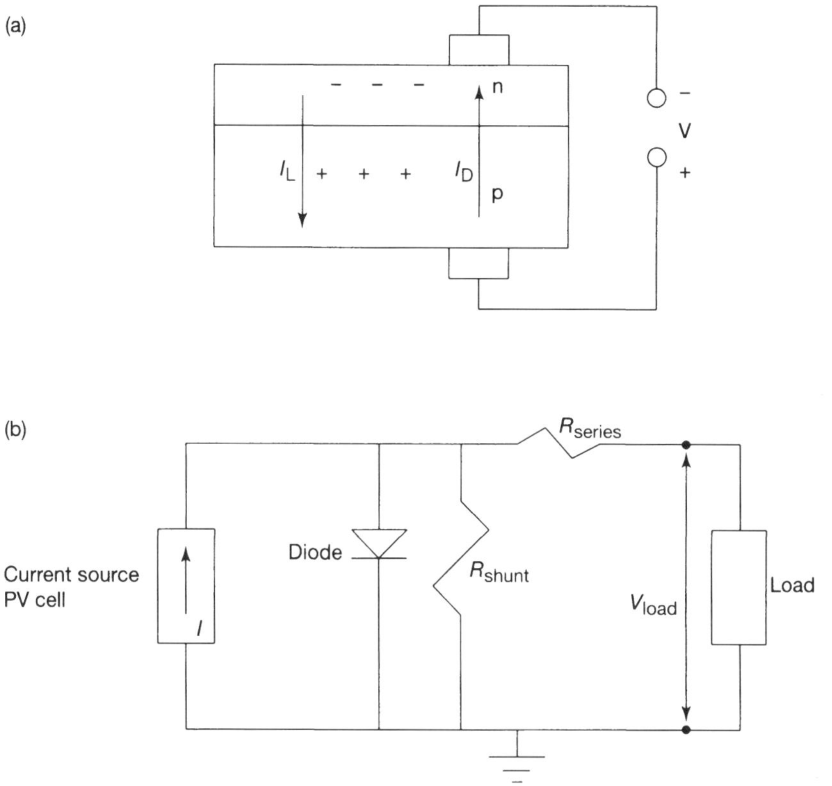

Figure 9.20 (a) Schematic of solar cell (+ represents holes and – electrons). (b) Equivalent circuit for PV cell and associated components

Gallium arsenide, which is much more expensive than silicon, also absorbs sunlight in a thin surface layer. Because its optimum operating temperature is higher than that of silicon, it is suitable for concentrating PV arrays where reflectors concentrate radiation onto the cells. It has a high potential efficiency of over 22%, with a theoretical maximum exceeding 30%. Another promising thin film is cadmium telluride with a record efficiency of about 14%.

Commercial modules usually consist of 30–40 cells in series to produce a voltage of about 12 V as required to charge batteries. Interconnections between cells in a module are usually made with flexible tinted copper strips soldered to the cells. The cells are usually encapsulated with a material such as ethylene vinyl acetate and covered with low-iron (high-transmittance) glass for weather protection. A frame is also included for mechanical strength and for easy mounting. An electrical junction box is usually attached to the back of the module, including the positive and negative leads; the junction box may also contain ground points, bypass diodes, etc. Crystalline modules are usually more reliable, with an expected minimum lifetime of 20 years.

SOLAR CELL EFFICIENCY

Part of the incident solar radiation is converted to electricity while the rest is reflected or converted to heat. An untreated surface reflects up to 35% of the incident solar radiation; surface coatings can reduce this to about 5%. Losses in the conversion of light to electricity occur because some of the sunlight (long wave) does not have enough energy to release electrons and passes through the semiconductor to be absorbed or reflected by the base of the electrode. Ultraviolet and very short wavelength light has more energy than required to release electrons and generates heat. These two opposing effects limit the maximum theoretical efficiency of single p-n junction cells to about 30%.

Other effects such as recombination of electrons and holes at defects reduces actual efficiency further. In addition, we have **resistance losses** due to the cell base material, the surface layers and connections; this is modelled by a **series resistance** $\boldsymbol{R_s}$ in the cell circuit (see Figure 9.20). Generally the efficiency also decreases with rise in temperature (about 0.5% per °C). We also have **shunt losses** because imperfect cell fabrication allows current shunt paths through the junction or around the edges. These losses reduce cell efficiency to between 5 and 30% (theoretical maximum).

The current I produced by the cell is approximately proportional to the solar irradiance. It also varies exponentially with voltage V (a function of load). At a fixed temperature and solar radiation, the voltage characteristic of the equivalent circuit in Figure 9.20b is given by

$$\begin{aligned} I &= I_L - I_D - I_{shunt} \\ &= I_L - I_o\{\exp[e(V + IR_{series})/(mkT)] - 1\} - (V + IR_{series})/R_{shunt} \end{aligned} \tag{9.18}$$

where I_D is diode current, e is the charge of an electron (1.602×10^{-19} C), k is the Boltzman constant (1.381×10^{-23} J/K), m is ideality factor (between 1 and 2) and T the cell temperature (K).

In a modern cell, the shunt resistance is usually high and I_{shunt} can be ignored. In addition, the term $\exp[e(V + IR_{series})/(mkT)]$ is much greater than 1. Finally, I_L is equal to the short-circuit current I_{sc}.

Setting $V_t = mkT/e$ (equal to about 25 mV at 300 K), equation 9.18 becomes:

$$I = I_{sc} - I_o\exp[(V + IR_{series})/V_t] \tag{9.18a}$$

This equation can be expressed in terms of open circuit voltage V_{oc} as

$$I = I_{sc}(1 - \exp[(V - V_{oc} + IR_{series})/V_t] \tag{9.18b}$$

For a PV module with N_s cells in series and N_p in parallel ($N_s \times N_p$ array) the above equation is modified as:

$$I = I_{scm} (1 - \exp[(V - V_{ocm} + I_m R_{sm})/(N_s V_t)] \tag{9.18c}$$

where the module parameters are:

$$I_{scm} = I_{sc} N_p$$
$$V_{ocm} = \mathrm{V}_{oc} N_s$$
$$R_{sm} = R_{series} N_s / N_p$$

A cell/module is characterized by its maximum power (peak watt rating of cell) given by:

$$P_{max} = V_{mp} \times I_{mp} \tag{9.19a}$$

under standard conditions of: $G = 1000$ W/m^2, $T = 25$°C (cell operating temperature).

The power output curve (Figure 9.21) is given by the power law:

$$P = V \times I \tag{9.19b}$$

where $V = I \times R$ (Ohm's law).

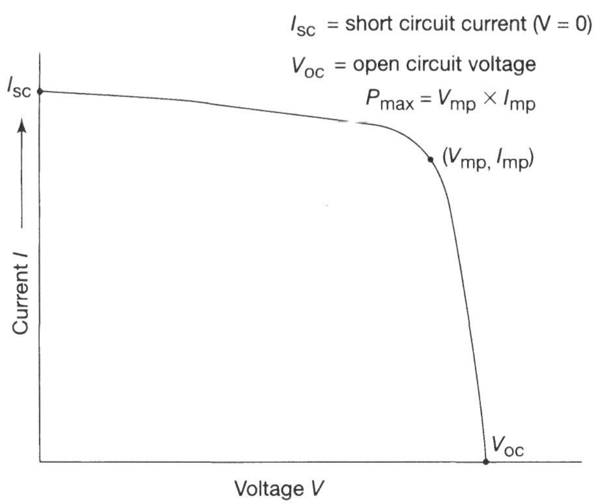

Figure 9.21 Variation of output for PV cell

When $V = 0$ we obtain the short-circuit current I_{sc}. As resistance increases, the voltage increases up to the open-circuit voltage V_{oc} (with $I = 0$).

A useful parameter for characterizing solar cells/modules is the **fill factor** FF defined by

$$FF = P_{max}/(V_{oc}I_{sc}) \tag{9.20}$$

Because V_{oc} and I_{sc} are the maximum possible voltage and current respectively, FF is always less than 1, and typically in the range 0.7–0.8 for most crystalline cells (Si, GaAs, etc.). The **energy conversion efficiency of a cell,** which is equal to electric power output divided by incident solar radiation (G_t), may be written as:

$$\eta_c = FFV_{oc}I_{sc}/(G_t) \tag{9.21}$$

Temperature effect

Usually, at high incident solar radiation the cell may operate at about 25°C higher than the ambient temperature due to the solar radiation absorbed and converted to heat. This rise in temperature reduces the cell efficiency by about 0.002 V/°C for each silicon cell. Thus, if a module operates at an ambient temperature of 25°C, its own temperature will be 50°C, and given that the PV module consists of 40 cells, we have a reduction in output voltage equal to:

$$DV = 0.002\ \text{V/°C/cell} \times 40\ \text{cells} \times (50°\text{C} - 25°\text{C}) = 2\ \text{V}$$

Therefore, if the operating temperature of the cell is reduced, its efficiency increases. Thus, it may often be advantageous to combine a thermal application with a PV application to remove some heat from the modules, cooling them and enhancing their efficiency. A typical application would be integration of PV modules into an airflow window (between two glazings). Air flows over the warm modules, absorbing useful heat and transporting it to a heated space.

The operating temperature T_c of the cell may be determined by performing an energy balance on the cell:

Absorbed solar radiation = electric power ouput + heat loss to environment

$$\tau\alpha G_t = \eta_c G_t + U(T_c - T_e) \tag{9.22}$$

where τ is cover solar transmittance, α is cell solar absorptance, U is the heat transfer coefficient to the environment and T_e is environment effective temperature; normally the product $\tau\alpha$ is about 0.9.

The nominal cell operating temperature (NCOT) is based on a solar radiation level of 800 W/m^2 (at normal incidence), T_e equal to 20°C and a wind speed of 1 m/s with no load (no output). From equation 9.22 we can see that:

$$\tau\alpha/U = (\text{NCOT} - T_e)/800 \tag{9.23a}$$

Since $\tau\alpha/U$ remains approximately constant, we can find the cell temperature at other than nominal conditions by rearranging equation (9.23a):

$$T_c = T_e + (G_t\tau\alpha/U)(1 - \eta_c/\tau\alpha) \tag{9.23b}$$

9.4.2 PV systems

A PV array consists of connected PV modules which deliver power to an external device such as a power conditioner. It contains interconnect cabling, a support structure and components such as blocking diodes, bypass diodes, test points, switches, junction boxes, fuses, reflectors, lightning protection systems and a grounding system. Some suppliers offer about four to ten modules assembled into a panel to reduce installation labour.

A bypass diode prevents a poorly performing (e.g. shaded or damaged) module from affecting other modules connected in series, by acting as a low-impedance bypass path. It maintains the current from the other modules and simply reduces the total output voltage by that of the poorly performing module. Bypass diodes may be integrated into modules. Blocking diodes prevent a reverse current from flowing from a source such as a battery into the PV modules.

SERIES AND PARALLEL CONNECTIONS

To charge a 36 V battery we need, for example, three standard 12 V modules connected in series. The number of modules in parallel is decided by the required total peak power output; the voltage is the same across each module as well as the current giving:

$$P_{\text{total}} = N \times V \times I$$

where N = number of modules in parallel.

If certain modules are shaded they should be bypassed. In an array with parallel groups of modules connected in series, shading of one module causes the bypass diode to activate, bypassing the affected group and taking it out of operation. Thus, arrays using series strings of PV modules connected in parallel are more common than parallel strings of modules connected in series (see Figure 9.22), as only a faulty module needs to be bypassed and not a whole group.

Grounding equipment protects from lightning, static charges, leakage currents from faulty components, etc. System grounding is also necessary; one of the conductors carrying current is ground at one point.

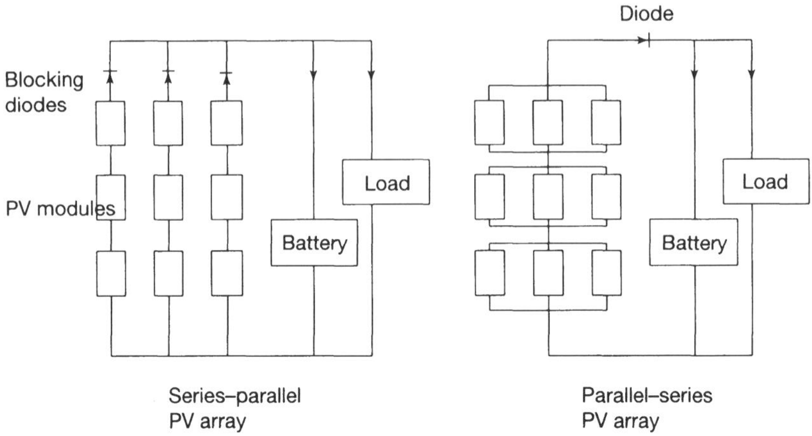

Figure 9.22 Typical PV array circuits

ENERGY STORAGE AND BATTERIES

There are various types of **lead–acid** batteries that store electricity as chemical energy such as:

- **simple lead–acid** with lead, lead oxide electrodes and sulphuric acid electrolyte
- **lead–antimony** batteries, in which antimony is added to the lead grid to provide extra strength; they can be discharged by 50–80% of capacity, but produce gas sometimes during recharge and lose water near full-charge; water must be added periodically
- **maintenance-free lead–calcium batteries** contain **calcium** on the lead grid which reduces gas formation and water loss; enough liquid is added to compensate for water loss during its estimated five-year life; its life is reduced if discharged to more than 25% of capacity.

Lead–acid battery cells have a nominal voltage of 2 V, which may vary from 1.75 V to 2.45 V. Six cells in series give a 12 V battery.

The **capacity** of a battery is given in **ampere-hours** under a temperature of 25°C and a charge rate of 8, 24 or 100 hours. The discharge rate is given in amperes or discharge time. The discharge time is the time required to discharge a battery completely at the discharge current.

Discharge rate (amps) = capacity (amp-hours)/discharge time (hours)

A C/10 rate means 10 hour discharge time and is the fastest possible without bubble formation (gassing). A C/20 rate is more acceptable, while for PV a C/100 rate is often assumed in design.

The **depth of discharge (DOD)** is the fraction of battery discharge before recharging; 20% is considered shallow discharge while 50–85% is deep. Battery voltage varies with DOD and specific gravity of electrolyte, as well as temperature. Battery capacity is often given at C/100.

Lead–antimony and nickel–cadmium batteries perform better at higher temperatures while lead–acid batteries have a decreasing charging voltage with rising temperature. If a battery is discharged deeply (e.g. more than 50%), its life is reduced; deep discharge also reduces the recharge voltage level and causes degradation of electrode separator material.

PV systems require batteries that can be charged during daytime and discharged at night, and to withstand many cycles of charge–discharge. Batteries' charge–discharge cycles can vary from 200 to 10 000 over their lifetime.

Days of autonomy (DOA) are the number of days for which the battery can supply electricity without recharge (and no damage).

The battery capacity C (amp-hours) required can be chosen based on the following equation:

$$C = L \times \text{DOA/DOD}$$

where

L = daily load in ampere-hours (A h)

DOD = depth of discharge per daily cycle

Systems designed to use 15–25% battery capacity each night are called **shallow cycle systems.** If we design for shallow discharge, battery lifetime is increased and lead–calcium batteries may be used. Remote telecommunication systems are designed for 30 days of autonomy (30 cloudy days) and a shallow discharge. In **deep cycle systems** the batteries are selected based on 50% or more DOD; in this case DOA is low.

Nickel–cadmium (NiCd) batteries are suitable for applications that require reliability, high DOA and low maintenance. They may last 20 years.

POWER CONDITIONING

Battery charge regulators switch off current from the PV array to the batteries when they are fully charged. Two-stage controllers allow regular charging during the first stage and trickle charge rate at the second stage permitting the voltage to increase to 18 V on a 12 V system; this trickle charging is good for most batteries. Battery charge regulators may also have a built-in blocking diode.

D.C.–D.C. converters change the voltage which varies as a function of insolation to a level required by the load. They are not very common in stand-alone systems with battery storage.

Inverters, which convert D.C. to A.C., are necessary when powering A.C. loads. In a system with a battery the inverter is connected to the battery. An inverter must have adequate surge capacity because inductive loads (such as motors) require a current surge for several seconds to start. The output voltage should also be regulated, so that as battery capacity drops, the output A.C. voltage does not vary by more than 5%. The inverter should not consume much energy, i.e. it should have an efficiency of at least 90%, so as not to significantly reduce overall system efficiency. **Square-wave** inverters switch the D.C. input into a square-wave A.C. output; they are suitable only for small resistive heating devices and lights, but not for motors. **Modified sine-wave inverters** use transistors or silicon-controlled rectifiers (SCR) to produce an approximate sine wave with low harmonic distortion.

A **maximum power point tracker** maintains the maximum possible power voltage independent of variations in insolation and load impedance; it is essentially an impedance matching device. An **array switcher** may be used to change the series/parallel configurations of modules in the array to achieve desired current level even though insolation level drops.

Load management controllers typically switch off loads from the battery in a sequential order as the battery voltage drops.

9.4.3 Load analysis in stand-alone systems

The electrical load is the sum of the individual power demands from each device connected to the system. Whenever possible, D.C. devices should be used because some power is lost in converting from D.C. to A.C. Fluorescent lighting fixtures with D.C. ballasts are suitable for PV systems. PV power is not very efficient for heating applications (except possibly small heat pumps). D.C. refrigerators are available.

SIZING METHOD

All-year systems should be sized based on the period with minimum insolation – December or January. Summer-only systems may be sized based on September or early October data. Typically, about 3–15 days of autonomy are desirable for a stand-alone systems.

- Identify weather and insolation conditions; find mean monthly insolation per day at the tilt angle of interest (three tilt angles may be considered – latitude, latitude plus 15° and latitude minus 15°) for each month; tabulate also mean ambient temperature.
- Estimate electrical load.
- Estimate required batteries to satisfy load for the required period of autonomy.
- Size array to charge batteries in one average day during the period with low insolation.

9.4.4 Building-integrated photovoltaic systems

One of the most promising applications of PV today is supplemental utility generation. A major benefit of PV is that it tends to generate more electricity at times of the year when more is needed. In most of North America, for example, utility companies struggle to keep millions of power-hungry air conditioners running in the summer months, when photovoltaic cells are most productive. Many utilities are using PV to help cover peak loads, rather than building entirely new plants. Because PV is so modular in design, these supplemental stations can easily grow from year to year, covering higher and higher loads. Several applications are described below.

BUILDING WITH PHOTOVOLTAICS (SCHOEN 1997)

Photovoltaics permit the on-site production of electricity without concern for fuel supplies or environmental adverse effects. Power is produced without noise and no depletion of resources. It is expected that in the 21st century building-integrated grid-connected PV systems will contribute significantly to power production. However, costs must still be lowered further. Cost reductions are achieved not only by increasing solar cell efficiency, but also by integrating the cells with the building envelope. In cases where PV facades replace expensive components such as granite or marble they can be cost-effective already, especially in the Mediterranean areas.

THE ARCHITECTURAL POTENTIAL OF THE GRATZEL CELL (FITZMAURICE 1997)

This is a transparent regenerative photoelectrochemical cell utilizing low to medium purity materials and simple construction processes, with a conversion 7–12%. While in conventional cells the semiconductor assumes the functions of both absorption and carrier transport, in the Gratzel cell light is harvested by dye molecules at the surface of the semiconductor film while charge transport is achieved by the nanostructured semiconductor film. The advantage of this cell is simplicity of production and low investment; the cost per kW is five times lower than that of opaque cells. Gratzel cells may be combined with conventional glazing by choosing a dye concentration such that the solar radiation transmitted is appropriate for interior daylighting.

In another approach, conventional and solar (with Gratzel cells) glazing may be combined.

THE INTEGRATION OF NEW TECHNOLOGIES IN BUILDINGS

PV cells have been integrated through various techniques into building envelopes:

- amorphous photovoltaic modules as a roofing material (Finland, Canada)
- insulated roof panels with amorphous PV modules on the top face (Japan)

- crystalline PV modules as atrium glazing – a skylighting system
- crystalline PV modules as a fixed shading device
- crystalline PV modules as wall cladding
- semitransparent amorphous PV cells as window panes (Germany).

PV systems need to be considered early in the building design process for their successful integration. Four successful applications are shown in Figures 9.23 and 9.24.

THE POMPEU FABRA LIBRARY MATARO BARCELONA (LLORET AND ACEVES 1997)
This is a 53 kW_p grid-connected building with integrated PV-thermal multi-functional modules (Figure 9.24, top). The building is the first of its kind in the European Union. Large-scale prefabricated multifunctional architectural modules producing both electrical and thermal solar energy were designed for industrial production. Various shapes and degrees of transparency were developed for curtain walls, roofs and blinds; solar thermal energy generated by solar absorption in the modules may be used for heating or evacuated via a ventilation chamber. The building is rectangular in shape with the largest facade facing south. The PV-thermal modules are the major innovation of the building, as they provide versatile and aesthetic possibilities for the building envelope while generating electricity and warm air which is fed into the gas heating system for further heating if necessary. The prefabricated PV-thermal modules include the complete electric connection system and are installed in the structure of the building using curtain wall construction methods. The building can produce about 125% of the calculated electricity needs.

9.5 Advanced fenestration system applications and integration issues

The building envelope may be designed as an energy-generating and storage system. Integration of photovoltaic systems is one possible option in this direction. However, there are also several other alternatives that integrate the concept of window–solar air collector with storage. Several examples are discussed below.

9.5.1 The new parliamentary building in London (23 000 m² floor area; six storeys)

Ventilated floor voids heat/cool structural slabs and enable displacement ventilation. The facade light shelves, which are also overhangs, reflect daylight to the back of the room. The floor bottom layer is high thermal capacity concrete which is the ceiling facing layer of the floor below; it was painted white to act as daylight reflector of the room below (Figure 9.25).

Figure 9.23 Use of PV panels on roof/solarium and as overhangs

Facade: the cladding system provides an integrated solution to the provision of view, daylight control, passive and active solar energy collection, shading, minimization of heat loss in winter, ventilation exhaust and supply and heat recovery. The fenestration consists of triple glazing with internal ventilated blinds; the outer part is double-glazed, argon-filled with low-emissivity coating. The inner cavity

Figure 9.24(a) Pompeu Fabra Library, Barcelona: utilization of semitransparent window systems with PV cells.

Figure 9.24(b) National Assoc. of Home Builders 21st century townhouse, Maryland, USA: effective roof integration of PV panels. © Bekaert ECD Solar Systems LLC., UNI-SOLAR® 2 kW of Solar Metal Roofing.

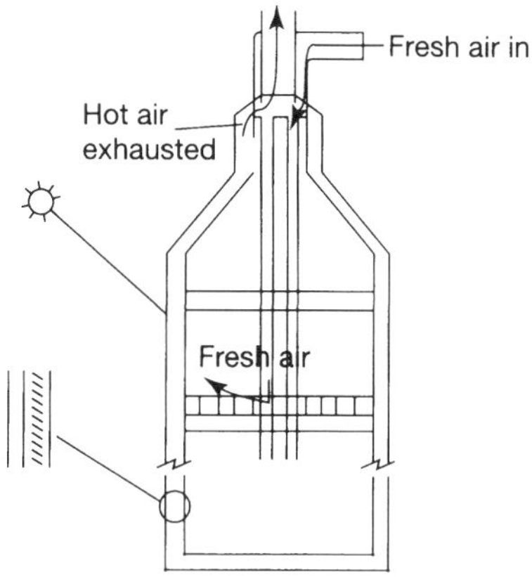

Figure 9.25 Schematic of new parliamentary building in London

contains retractable dark louvre blinds to maximize the absorbed solar heat. This cavity is ventilated with a proportion of room exhaust air and acts as solar collector. The window arrangement uses a light shelf to maintain room daylight when blinds are closed, thus avoiding the commonly observed case of blinds closed and lights on, with its consequent heat gain due to luminaires and extra cooling load. The facade is a passive system – only the blinds move.

Cooling: solar heat collected during daytime is stored for possible night use. When the outside air is above 19°C, groundwater at 14°C is drawn from two on-site boreholes to cool the outside supply air to room temperature. Night ventilation is used for mass precooling.

Ventilation: the mechanical ventilation system serves a network of floor plena linked via a ductwork in the facade to provide 100% outside air ventilation to each room. High-efficiency heat recovery is the key benefit of mechanical ventilation over natural ventilation (rotary heat exhangers at 85% effectiveness; can also be hygroscopic to recover winter moisture). Low-pressure-loss air-handling equipment is employed – energy use of 1 W per litre/s of air supplied; total fan pressure 640 Pa with fan efficiencies 65%; typical air handling unit face velocities 1.2 m/s with filters at 0.8 m/s.

Heating: condensing natural gas boilers, variable-water-volume system with thermostatic valves on room heat emitters; water at 70°C supply and 50°C return; mass of water flow is halved compared with practice and using piping at 50 Pa/m pressure loss results in significant pump power reduction (90 kWh/m^2 annual energy consumption – possibly 65 in optimized design).

9.5.2 New tax office (Enschede, Netherlands)

The project was part of the THERMIE project 'Energy Comfort 2000', a group of eight innovative projects in Europe. It is 4300 m^2 and five storeys. Natural ventilation was employed – a central atrium was used for extraction of air. Fresh air enters by means of ventilation grills independent of wind pressure and situated near the workstations on the facades.

Daylighting: two strips of window separated by a light shelf are used for each storey; the lower part provides a view and the higher part admits extra daylight. The light shelf reduces glare and reflects some daylight onto the ceiling and then deep into the room. Venetian blinds are used on the outside of the lower window strip; a central control system closes the blinds in the morning when necessary, while during the day they are operated manually by the user.

High-frequency lamps are used with automatic dimmers capable of lowering emission to 30% when daylight is plentiful.

9.5.3 Thermotropic materials and systems for overheating protection

Thermotropic materials that change from transparent to white and vice versa as a function of temperature may contribute to the reduction of room overheating by preventing unwanted solar radiation from entering the room when the temperature is high. Thermotropic systems consist of at least two components with different refractive indices. At low temperatures these are mixed at the molecular level so that the material is transparent. If the material temperature rises to a certain level (depends on the mixture ratios) the two components separate into microscopic domains and incident radiation is scattered, most being diffusely reflected, and the material appears white.

Two types of materials are investigated at the Fraunhofer ISE – hydrogels and polymer blends. Hydrogel layers have good optical properties and low switching temperatures (20–50°C), which makes them suitable for glazings (see Figure 9.26). Polymer blends have switching temperatures of 30–120°C and they are thus suitable for transparent insulation systems. Thermotropic hydrogels and polymer blends have been produced in areas of more than a square metre with industrial technology; they switch solar energy transmittance by 55%. More research is being carried out on their further improvement and durability, and to achieve a steeper switching gradient.

9.5.4 Intelligent control of window shading systems

Software is available that can evaluate energy benefits from the use of various shading systems and daylight control devices. However, it is important to utilize the potential of today's smart digital controllers and building automation systems

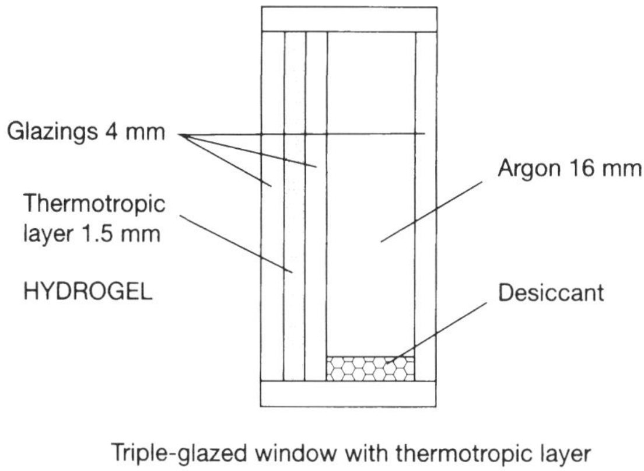

Figure 9.26 Use of thermotropic material in a window

(BAS) to optimally control the level of daylight and its quality (reduction of glare), while reducing cooling loads and heating loads. For example, they may be employed to dim electrical lights as a function of available daylight, while choosing an optimal position for motorized shading devices to reduce heating and cooling energy consumption.

Effective utilization of fenestration with motorized shading devices, preferably operated through a building automation system, requires two major design objectives which address the daylighting and energy aspects as follows:

- First, there is a need to **control the quantity and direction of the daylight transmitted**, especially in an office area, so as to minimize the need for electric lighting while avoiding glare problems.
- **Overall energy costs for electric lighting, heating and cooling need to be minimized**. If the first objective above is achieved, then the electric lighting costs are reduced and cooling loads are lowered because of avoidance of heat emission from light fixtures. Ideally, if the available daylight does not meet the required illuminance levels in the work area, the intensity of electric lights should be variable (dimmable) to provide the additional required illuminance. However, if more than enough daylight is available during the cooling season, then the excess should be excluded in order to prevent increase of cooling loads. Finally, there is one more case to be considered: if passive solar heating is desirable, then the amount of total solar radiation transmitted into a zone should be maximized.

Figure 9.27 shows the schematic of an outdoor test-room (at Concordia University in Montreal) with a window having motor-operated reflective blinds between two glazings. These blinds are operated by a computer, which also controls the heating/cooling system and electrical lights.

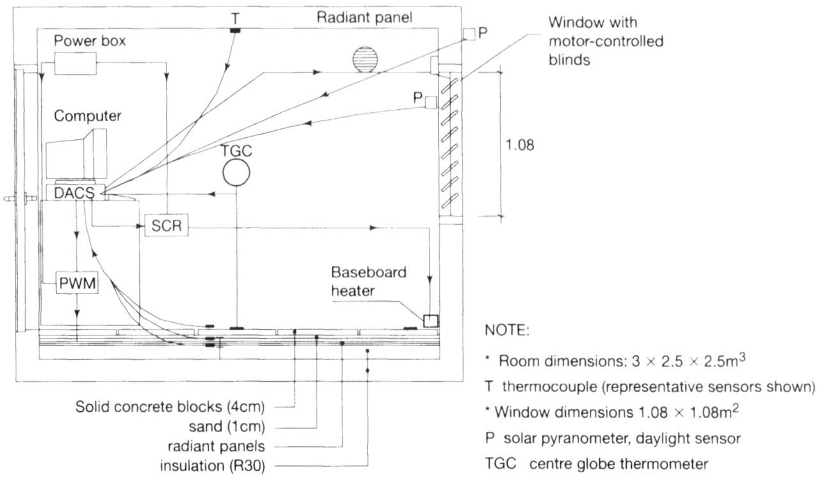

Figure 9.27 Schematic of outdoor test-room with computer controlling venetian blinds integrated in glazing, and heating system; DACS, data acquisition system; PWM, pulse-width modulation; SCR, silicon-controlled rectifier

The focus of the project is on development of control techniques and software for operation of shading devices, in conjunction with dimming of electrical lights and heating/cooling control so as to optimize building energy efficiency and occupant comfort. Its specific objectives are the following:

- Development of control procedures and algorithms for operation of motorized window shading devices, so as to transmit an **optimum amount of daylight** while reducing glare problems. Input variables include exterior incident daylight, and the control variable is the blind tilt angle or roller shade position.
- Development of control procedures and algorithms for operation of motorized shading devices in conjunction with control of electrical lights and heating/cooling control in order to **enhance overall energy efficiency** (minimize heating and cooling energy consumption, as well as electricity consumption for lighting).
- Implementation of the above control algorithms in a BAS (computerized energy management and control system).
- Evaluation of the potential improvement in energy efficiency and luminous environment due to implementation of the above algorithms.

The digital control algorithms developed (Athienitis and Tzempelikos 2001) may be implemented in local smart controllers operating the window shading systems or in a central automation system (BAS). Figure 9.28 shows typical results from implementation of the algorithm for a typical office with windows having motorized blinds. A blind tilt angle is selected to prevent beam solar radiation from directly entering the workspace thus preventing glare.

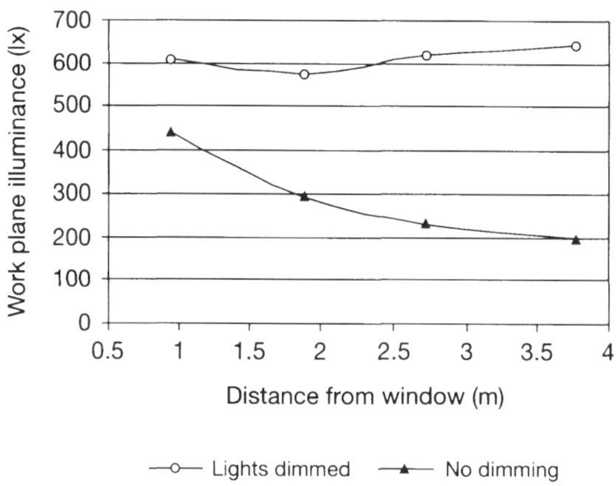

Figure 9.28 Work plane illuminance as a function of distance from window for typical 5 m × 5 m office with daylight only (no dimming) and with dimmed lights plus daylight (Montreal latitude 45°N, clear day at 9 a.m. in December)

Morel *et al.* (1996) developed a fuzzy logic controller to decide roller blind position. The objective of this project at LESO was the development of an optimal blind controller, taking into account optimization of thermal comfort, daylighting, minimization of energy consumption (thermal and lighting) and user wishes. Fuzzy logic was employed. Control algorithms were tested with simulation. The controller was installed in an office with high passive solar gains.

Optimization of blind position takes into account both daylighting and artificial lighting as well as passive solar gains. The control algorithm takes into account passive solar gains, internal gains etc. using solar sensor, electric power sensor and infrared occupancy sensor.

9.5.5 Integrated facade concept

Figure 9.29 illustrates a perimeter office in a commercial building which integrates several 'green' renewable energy technologies:

- photovoltaic panels integrated on the opaque (spandrel) section of the curtain wall
- a vent for natural ventilation
- top window with motorized integrated reflective blinds for reflecting daylight into adjacent corridor and interior zone of the building
- middle window with roller blind.

An important characteristic of the design is the requirement for a high ceiling, and two thirds of the facade glass (vision area) to maximize usable daylight. The

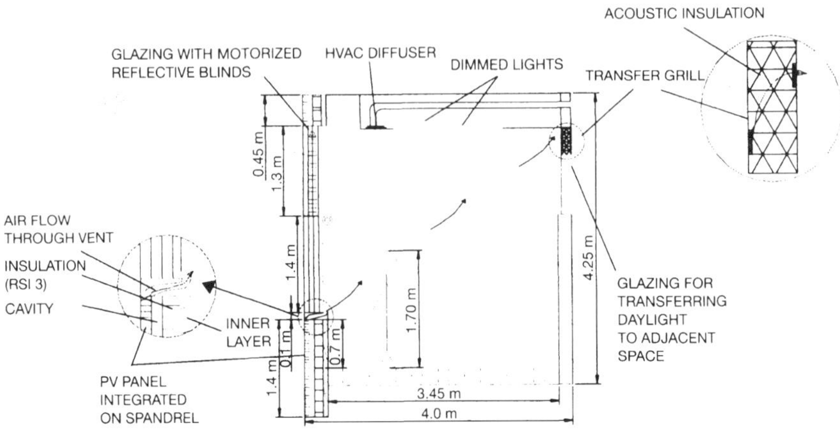

Figure 9.29 Perimeter office with facade-integrated photovoltaic panels, vent and motorized reflective blinds

transfer grill serves to transfer air coming through the vent and may or may not be necessary, depending on the natural/hybrid ventilation concept for the building.

The ideal type of glazing for maximization of daylight penetration in the building is clear glass. This requires careful selection of shading devices with optimal transmittance and low absorptance. High solar gains and natural ventilation through the integrated vent make the presence of exposed thermal mass essential in minimizing room temperature fluctuations. Therefore, in such a perimeter office with an acoustic ceiling, it is essential to avoid the use of carpets which insulate floor thermal mass. Instead, it is preferable to have a concrete or tile floor, which is also better than a carpet for good indoor air quality. Using the techniques introduced in previous chapters it may be shown that the room admittance is much lower with a carpeted floor than without the carpet.

9.5.6 Conclusion

Development of new technologies such as small-scale fuel cells, combined heat and power (CHP) cogeneration devices, lowering of the costs of photovoltaic systems integrated in buildings, and the adoption of advanced fenestration systems, will revolutionize building design in the twenty-first century. In certain climates, a building may eventually become a net energy producer as opposed to an energy consumer.

The focus on a high-quality indoor environment, both at home and in the workplace, will be the primary objective of building design, with energy efficiency being the essential constraint.

These changes will radically affect the way we live and the way we transport ourselves. Electric or hybrid cars will inevitably get a boost from an electric power producing building. Environmental concerns, which correctly call for urgent measures to reduce carbon dioxide emissions, coupled with rising oil prices, have created a renewed interest in electric and hybrid cars, as well as in solar buildings. The coming revolution or evolution in building design and operation will inevitably cause significant changes in the way cities and transportation systems are designed and built.

There is one essential requirement in the development of better buildings: building design and operation planning will have to be carried out in an integrated manner, with architects and engineers working together towards the creation of an energy efficient high-quality indoor environment.

9.6 Bibliography

Arulanantham M. and Kauskika N.D. (1994) 'Global radiation transmittance of transparent insulation material', *Solar Energy*, **53**:4, 323–328.

ASHRAE (1989) *ASHRAE Handbook – Fundamentals*, Atlanta, GA.

Athienitis A.K. (1988) 'A predictive control algorithm for massive buildings', *ASHRAE Transactions*, **94**:2, 1050–1068.

Athienitis A.K. (1994) *Building Thermal Analysis*, Electronic Mathcad Book, MathSoft Inc., Boston, MA.

Athienitis A.K. (1997) 'Theoretical investigation of thermal performance of a passive solar building with floor radiant heating', *Solar Energy*, **61**:5, 337–345.

Athienitis A.K. (1998) *Building Thermal Analysis*, 2nd edn, Electronic Mathcad Book, MathSoft Inc., Boston, MA.

Athienitis A.K. and Chen T. (1993) 'Experimental and theoretical investigation of floor heating with thermal storage', *ASHRAE Transactions*, **99**:1, 1049–1057.

Athienitis A.K. and Shou J. (1991) 'Control of radiant heating based on the operative temperature', *ASHRAE Transactions*, **97**:2, 787–784.

Athienitis A.K. and Tzempelikos A. (2001) 'A methodology for detailed calculation of room illuminance levels and light dimming in a room with motorized blinds integrated in an advanced window', eSim2001 Building Simulation Conference, Ottawa, Canada, June.

Boonstra C. and Vollebregt R. (1997) 'Space for architecture in passive office buildings', *European Directory of Sustainable & Energy Efficient Building*, James and James Ltd, Hong Kong, 25–28.

Brandemuehl M.J. and Beckman W.A. (1980) 'Transmission of diffuse radiation through CPC and flat plate collectors glazing', *Solar Energy*, **24**, 511–513.

Cane R.L.D., Hollands K.G.T., Raithy G.D and Unny T.E. (1977) 'Free convection heat transfer across inclined honeycombs panels', *ASME Transactions*, **Feb.**, 86–91.

Charters W.W.S. and Peterson L.F. (1972) 'Free convection suppressing using honeycomb cellular materials', *Solar Energy*, **13**, 353–361.

Dunster B. and Pringle J. (1997) 'Sustainable architecture and the low energy urban office', *European Directory of Sustainable & Energy Efficient Building*, James and James Ltd, Hong Kong, 16–24.

Eicker U. and Fux V. (1999) 'Thermal performance of building-integrated ventilated PV facades', Proc. ISES99 World Solar Congress, Jerusalem.

Elsherbiny S.M., Raithby G.D. and Hollands K.G.T. (1982) 'Heat transfer by natural convection across vertical and inclined air layers', *ASME Transactions*, **104**:1, 96–102.

Feldman D., Banu D., Hawes D. and Ghanbari E. (1991) 'Obtaining an energy storing building material by direct incorporation of an organic phase change material in gypsum wallboard', *Solar Energy Materials*, **22**, 231–242.

Feldman D., Khan M.A. and Banu D. (1989a) 'Energy storage composite with an organic phase change material', *Solar Energy Materials Journal*, **18**, 333–341.

Feldman D., Shapiro M., Banu D. and Fuks C.J. (1989b) 'Fatty acids and their mixtures as phase change materials for thermal energy storage', *Solar Energy Materials Journal*, **18**, 201–216.

Fitzmaurice D. (1997) 'The architectural potential of the Gratzel cell', *European Directory of Sustainable & Energy Efficient Building*, James and James Ltd, Hong Kong, 84–85.

Francia G. (1961) 'Un nouveau collecteur de l'energie rayonnante solaire – theorie et verifications experimentales', *Conference des Nation Unies sur les sources nouvelles d'energie*, E/conf.35/S/71, 554–558.

Georg A., Graf W., Schweiner D., Wittwer V., Nitz P. and Wilson H. (1998) 'Switchable glazing with a large dynamic range in total solar energy transmittance (TSET)', *Solar Energy*, **62**:3, 215–228.

Goetzbeger A., Schmid J. and Wittwer V. (1984) 'Transparent insulation system for passive solar energy utilization in buildings', *International Journal of Solar Energy*, **2**, 289–308.

Hawes D.W. (1991) 'Latent heat storage in concrete', Ph.D. Thesis, Concordia University, Montreal, Canada.

Hollands K.G.T. (1965) 'Honeycomb devices in flat-plate collectors', *Solar Energy*, **9**, 159.

Hollands K.G.T and Iyankaran K. (1985) 'Proposal for a compound honeycomb collector', *Solar Energy*, **34**:4, 309–316.

Hollands K.G.T., Marshall K.N. and Wedel R.K. (1978) 'An approximate equation for predicting the solar transmittance of transparent honeycomb', *Solar Energy*, **21**, 231–236.

Hollands K.G.T., Raithby G.D. and Unny T.E. (1976) *Studies on Methods of Reducing Heat Losses from Flat Plate Solar Collectors, Phase I Report*, University of Waterloo.

Holman J.P. (1986) *Heat Transfer*, McGraw-Hill Inc., New York.

Hottel H.C. (1976) 'A simple model of transmittance of direct solar radiation through clear atmospheres', *Solar Energy*, **18**.

Kaushika N.D. and Padma Priya R. (1991) 'Solar transmittance of honeycomb and parallel slat arrays', *Energy Conversion Management*, **32**:4, 345–351.

Kedl R. (1991) 'Wallboard with latent heat storage for passive solar application', ORNL/TM-11541 Distribution Category UC-202.

Lloret A. and Aceves O. (1997) 'The Pompeu Fabra Library Mataro Barcelona', *European Directory of Sustainable & Energy Efficient Building*, James and James Ltd, Hong Kong, 92–95.

Lorenzo E. (1994) *Solar Electricity*, Progensa, Sevilla, Spain.

Morel N. *et al.* (1996) *Delta: A Blind Controller Using Fuzzy Logic*, Final Report, LESO, Switzerland.

Olsen L. (1982) 'Transparent insulation for thermal storage walls', *Solar Energy R & D in the European Community Series A*, **2**, 127.

Őzisik M.N. and Uzzell J.C. (1979) 'Exact solution for freezing in cylindrical symmetry with extended freezing temperature range', *Journal of Heat Transfer*, **101**, 331–334.
Platzer W.J. (1992) 'Total heat transport data for plastic honeycomb-type structures', *Solar Energy*, **49**:5, 351–358.
Platzer,W.J. (1994) 'Transparent insulation materials: a review', *SPIE Conf. Proc.* 2255, Optical Materials Technologies for Energy Efficiency and Solar Energy XIII, 616–627.
Ramadan H. (1998) 'Numerical simulation model of a test-room with transparent insulation', MASc. thesis Building Engineering, Concordia University, Montreal, Canada.
Ramadan H. and Athienitis A.K. (1998) 'Numerical simulation of a building with transparent insulation', *Proceedings of the Solar Energy Society of Canada*, 24th Annual Conference, Montreal, May.
Santamouris M. *et al.* (1994) *Passive Cooling of Buildings*, CIENE, University of Athens.
Schoen T. (1997) 'Building with photovoltaics', *European Directory of Sustainable & Energy Efficient Building*, James and James Ltd, Hong Kong, 78–83.
Tabor H. (1969) 'Cellular insulation (honeycombs)', *Solar Energy*, **12**, 549–552.
Trombe F., Robert J.F., Cabonot M. and Sesolis B. (1977) 'Concrete walls to collect and hold heat', *Solar Age*, **2**, 13.
Twidell J.W., Johnstone C., Zuhdy B. and Scott A. (1994) 'Strathclyde University's passive solar, low energy, residences with transparent insulation', *Solar Energy*, **52**:1, 85–109.
Wilson H.R., Raicu A. and Nitz P. (1997) 'Thermotropic materials and systems for overheating protection', *European Directory of Sustainable & Energy Efficient Building*, James and James Ltd, Hong Kong, 63–66.
Zmeureanu R. and Fazio P. (1988) 'Thermal performance of a hollow core concrete floor system for passive cooling', *Building and Environment*, **23**:3, 243–252.

Index

Made in the USA
Monee, IL
13 September 2026

62978973R10164